REMOTE SENSING AND GEOGRAPHIC INFORMATION SYSTEMS FOR ENVIRONMENT AND NATURAL RESOURCES MANAGEMENT

Proceedings of the Workshop on Sustainable Rural Development Using Integrated Remote Sensing and GIS
Hyderabad, India, 16 to 20 September 1996

Jointly organized by
the Department of Space, Government of India
and the Economic and Social Commission for Asia and the Pacific

UNITED NATIONS
New York, 1997

Front Cover: Weekly composite AVHRR image of sea surface temperature of north Indian Ocean.

ST/ESCAP/1789

UNITED NATIONS PUBLICATION

Sales No. E.97.II.F.29

ISBN 92-1-119765-1

The designations employed and the presentation of the material in this publication do not imply the expression of any opinion whatsoever on the part of the Secretariat of the United Nations concerning the legal status of any country, territory, city or area, or of its authorities, or concerning the delimitation of its frontiers or boundaries.

Mention of firm names and commercial products does not imply the endorsement of the United Nations.

Any part of this publication may be reprinted if the source is indicated. However, if an item is republished in another language, the United Nations does not accept responsibility for the accuracy of the translation.

The views expressed in signed articles are those of the authors and do not necessarily reflect those of the United Nations.

This document has been issued without formal editing.

TECHNICAL EDITOR'S NOTE

In editing this report, every effort has been made to retain the original contents as presented by the authors. Editorial adjustments have sometimes been made in contents and/or illustrations which do not seriously affect the original form of the presentations.

CONTENTS

Page

PART ONE
REPORT OF THE WORKSHOP

PART TWO
OVERVIEW OF REMOTE SENSING AND GEOGRAPHIC INFORMATION SYSTEMS

PART THREE
SUSTAINABLE RURAL DEVELOPMENT PLANNING

PART FOUR
LAND USE/LAND COVER

CONTENTS *(continued)*

PART ONE
REPORT OF THE WORKSHOP

REPORT OF THE WORKSHOP ON SUSTAINABLE RURAL DEVELOPMENT USING INTEGRATED REMOTE SENSING AND GIS

A. Organization of the workshop

1. The Workshop on Sustainable Rural Development Using Integrated Remote Sensing and GIS was held in Hyderabad, India, from 16 to 20 September 1996. The workshop was organized by the Economic and Social Commission for Asia and the Pacific (ESCAP) in cooperation with the National Remote Sensing Agency (NRSA), Department of Space, Government of India. The main objective of the workshop was to exchange technical information on the use of integrated spatial information technologies for different types of rural development projects.

1. Attendance

2. The workshop was attended by 36 participants from 16 countries and organizations: Azerbaijan, Bangladesh, Cambodia, China, India, Indonesia, Islamic Republic of Iran, Malaysia, Mongolia, Myanmar, Nepal, Republic of Korea, Sri Lanka, Thailand, Viet Nam and ESCAP. Six observers from India also attended the Workshop.

2. Opening of the workshop

3. An address of welcome was delivered by Professor B.L. Deekshatulu, Director, National Remote Sensing Agency. The message of the Executive Secretary of ESCAP was read by an officer of the Space Technology Applications Section, Environment and Natural Resources Management Division of ESCAP. A keynote address was given by Mr George Joseph, Director, Space Applications Centre. The workshop was officially inaugurated by Mr Shri S.K. Akora, Director, National Institute of Agricultural Extension Management, and this was followed by a vote of thanks by Mr D.P. Rao, Associate Director, National Remote Sensing Agency.

4. In his welcome address, Professor Deekshatulu remarked that human creativity and ingenuity could solve the problems created by humankind. Technologies such as remote sensing and GIS needed to be urgently used by developing countries for making suitable development plans. In that context, he stressed the importance of human resource development.

5. In his message to the participants, the Executive Secretary of ESCAP noted that remote sensing and GIS technologies had become essential tools for development planners. The advantages offered by the integrated uses of that technology included greater speed and efficiency, innumerable analytical possibilities and the capacity to evaluate various possible development scenarios before they were actually implemented.

6. In his keynote address, Mr Joseph introduced the Indian space development programme which included the INSAT programme, the GRAMSAT concept and the remote sensing programme. He stated that remote sensing data was extremely useful and could be incorporated into many different areas of work. In that regard, he said that although the Indian space programme so far covered many areas of communication, meteorology, remote sensing and so on, the integrated technology of various fields had become more important and should be utilized for sustainable rural development.

7. In his address, Mr Shri S.K. Akora emphasized that any effort at rural development had to take into account the human element for the micro-level planning of watersheds. He believed that remote sensing technology was becoming increasingly more accurate and was making it possible to do the necessary inventories of natural resources at the village level.

3. Election of officers

8. Mr Herath Manthrithilake, Director, Mahaweli Environment and Forest Conservation Division, Mahaweli Authority of Sri Lanka was elected Chairperson; Mr Pradip P. Upadhyay, Section Officer, National Planning Commission Secretariat of Nepal, Vice-Chairperson; and Mr Loh Kok Fook, Head, Applications and Image Processing Division, Malaysian Centre for Remote Sensing, Rapporteur.

B. Workshop proceedings

1. Host country presentations

9. Several presentations were given by experts from various departments of the Government of India concerned with different aspects of sustainable rural development which integrated the use of remote sensing and GIS.

(a) Overview of integrated remote sensing and GIS for rural development in India

10. India was one of the few developing countries to have its own indigenous operational space and ground segment as well as the necessary expertise and infrastructure for operationally using remote sensing techniques in projects related to rural development. With its agriculture-based economy, the country had given top priority to the management of water resources, including the prioritization of watersheds, monitoring surface water bodies and groundwater management, snow-melt run-off and identification of suitable sites for water harvesting. The mapping of wastelands had also been taken up for critical areas at the national level using satellite images at 1:50,000 scale. Through the Integrated Mission for Sustainable Development (IMSD) taken up by the Department of Space of India, GIS had been used to integrate spatial data on various resource themes and socio-economic aspects and produce alternate development plans showing site-specific primary production activities. After presenting the results of pilot studies carried out in several drought-affected districts to the Central Planning Commission, a major programme was initiated to cover all districts with the active involvement of various national institutions working in different disciplines.

11. National resources represented a great asset in the economy of any country. Large reserves of resources were still unexplored and underutilized. Serious ecological problems might result from sector-oriented and independent exploitation. An integrated approach to the exploitation, conservation and management of natural resources and the protection and preservation of the environment was found to be essential. The integration of geographic information systems (GIS) and image analysis technologies was shown to play an important role in development planning, implementation and monitoring and in assisting in the exploration, exploitation and judicious management of natural resources.

12. Basic definitions and concepts in GIS and related technologies were presented and current issues and potential areas of application and research and development areas were discussed. The development of the Natural Resource Information System (NRIS) was also described.

13. The approach and methodology of the Integrated Mission for Sustainable Development (IMSD) were described. IMSD aimed at fulfilling basic needs, promoting economic growth and ecological balance, and was operationalized through a software package (Geo-LAWNS) designed and developed by the Department of Space, for integrated land and water resource planning for rural development. The package supported operationalized methodologies for generating thematic information on natural resources, using satellite data interpretation by visual and digital techniques, and allowed the integration of available collateral information on climate, demography and socio-economic aspects, data analysis through appropriate spectral analysis functions and the development of pragmatic developmental plans, consistent with resource potentials and problems, basic needs of the people, government priorities and national policies for sustainable development of the region.

14. The planning of development activities, especially in rural India, was also presented. The methods to devise sustainable development schemes used remote sensing data and conventional data

such as Survey of India maps, district books, census data and National Informatics Centre (NIC) data. For land resource management, geomorphology, existing land use, soil, surface water and groundwater were mapped at 1:50,000 scale using remotely sensed data. Conventional data included rainfall, transport network, village location and extent, watershed and drainage information, slope and social parameters such as demographic details, literacy levels and so on. Integration of that information helped to produce suitable action plans for the sustainable development of each land parcel unit in a watershed.

15. Detailed micro-level planning at the village level to more scientifically combat drought on a long-term basis, using remote sensing techniques in a time- and cost-effective basis, was presented. Basic information derived from remotely sensed data and conventional techniques about land and water resources at 1:50,000 scale were integrated and mapped. Keeping in view the present-day cropping patterns and the needs of the people, the methodology adopted results in Recommended Optimal Land Use and Farming Systems (ROLUFS). Various drought-proofing works such as optimal *in situ* soil and moisture conservation measures, rainwater harvesting structures, soil and moisture conservation measures, fodder, fuel wood and permanent tree cover development zones could be recommended and later validated in the field through participatory rural appraisal (PRA) exercises. The programme was implemented through a watershed development committee comprising the beneficiaries themselves with the assistance of non-governmental organizations (NGOs).

(b) Forestry and land degradation applications

16. Rapid depletion of forest resources and their biodiversity in the tropics had necessitated improved land use planning and classification. Remote sensing technology played a key role in surveying, assessing the resources and recording the changes. Satellite data could also help in developing biodiversity management. The integration of ground measurements with remote sensing data through GIS could provide information over large areas about environmental changes, while models could make future predictions about the environment. The current state-of-the-art remote sensing technology could meet most of the forest management requirements at various levels.

17. The new generation of satellites including IRS-1C and IRS-P3, had enhanced the scope for the monitoring and evaluation of forest resources. The assessment of fuelwood non-forest timber resources, non-timber resources and grazing resources as well as carrying capacity studies, the development of catchment treatment plans, forest hydrology, the monitoring of afforestation programmes, and the use of panchromatic data for micro-level planning were discussed. The five-day repeat cycle of IRS-1C WiFS data and its utility for forest fire monitoring and preventive measures were also emphasized.

18. The indiscriminate use of land for meeting the growing demand for food, fuel and fodder of an ever-growing population had led to serious environmental problems. By providing a synoptic view of large areas in discrete bands of the electromagnetic spectrum at regular intervals, space-based spectral measurements constituted an ideal tool for generating information on the extent, spatial distribution and temporal behaviour of degraded lands, including eroded lands, salt-affected soils, waterlogging, shifting cultivation and mining areas. The utility of various remotely sensed data, namely from the Landsat, SPOT and IRS-1A and 1B satellites in delineating, mapping and monitoring degraded lands was demonstrated through a few typical case studies.

(c) Land use management applications

19. Land use and land cover mapping and management using satellite data were presented. The existing (conventional) scenario in India for data collection, its limitations and problems were also discussed. The detailed methodology and the results obtained from national-level projects to map land use and land cover, wastelands and urban sprawl at different scales using satellite data were presented. The capabilities of IRS-1C data in analysing the micro-level land use and land cover at a large scale were also covered.

20. In view of the competing demands of the land for industries, human settlement, irrigation and infrastructure projects, it had been necessary to adopt land use planning on a scientific basis and better resource management techniques so that productivity and income improved without degrading the environment. Some of the measures required in that direction included (a) reclamation/development of wastelands, (b) proper allocation of land amongst different productive and consumptive uses, (c) periodic evaluation of the dynamic components of land like soil health, vegetation status and water use, and corrective management actions to bring stability in land use and (d) regulation of development activities through impact assessment models so that consequences were predicted and only proper actions were executed.

(d) Water resource management and fisheries applications

21. The importance, need, and problems associated with water resource management in developing countries were highlighted. Remote sensing applications in water resource management discussed included the estimation of precipitation, snow-melt and river run-off, surface and sub-surface storage as well as the management of irrigation water, watersheds and natural disasters such as floods, cyclones and droughts and water quality monitoring and management. Case studies were presented and it was concluded that even though considerable progress had been made in water resource management applications by remote sensing, sustained efforts were still needed for the development of operational packages and decision support systems.

22. Groundwater was one of the most important resources in the context of sustainable rural development. In India, more than 90 per cent of rural and nearly 30 per cent of urban drinking water supply was from groundwater. It also accounted for nearly 60 per cent of the total irrigation potential in the country. The geology, geomorphology, geological structures and recharge conditions control the movement and occurrence of groundwater. If precise information about all those factors was available, it was possible to infer the groundwater regime, i.e. the type of aquifer, type of wells suitable, depth range, yield range, success rate, area of influence of wells and their sustainability. Satellite data provided integrated information on all those factors and helped in groundwater resource estimation, budgeting, identification, mapping of prospective groundwater zones, systematic planning and development, augmentation and conservation of the resource and so on. Case studies were presented illustrating the use of satellite data in all the above areas. The methodology adopted for preparing district-wise hydro-geomorphological maps at 1:250,000 scale was explained, as well as ongoing research in the field using high-resolution IRS-1C satellite data. With multispectral LISS-III data (23.5 m resolution) and panchromatic data (5.8 m resolution), site-specific studies could also be carried out to solve the drinking water requirements of problem villages and the industrial and agricultural needs of the individual consumer/farmer within their land holding.

23. Based on sea surface temperature (SST) images, thermal features such as eddies, current boundaries and thermal gradients could be recognized, and forecasts indicating the potential fishing zones (PFZ) had been produced regularly in Hyderabad. Daily SST images of three or four days were composited and the minimum and maximum of the relative thermal gradient features such as thermal boundaries, fine scale relative temperature gradients to sharp contour zones, front eddies and upwelling zones were identified. Location of the PFZ with reference to a particular fishing centre was drawn by identifying the nearest point of the thermal feature to that fishing centre. The PFZ maps were sent through facsimile transmission to major fishermen associations, unions and concerned governmental organizations and state fisheries departments of all the maritime states every Monday and Thursday.

(e) Agriculture, soils mapping and related applications

24. Timely information on crop acreage and production estimates was essential for making strategic decisions for agricultural planning, import-export negotiations and storage, pricing and distribution of agricultural commodities. The conventional methods of obtaining such information were time-consuming and tedious. Satellite remote sensing offered a special advantage because of its ability

to provide synoptic coverage of large areas at regular intervals. The basic information that could be provided by remote sensing techniques for agricultural purposes included crop identification, crop acreage estimation, crop condition and crop yield forecasting. Procedures developed for pre-harvest acreage estimation of major crops using field sampling and digital techniques had been successfully operationalized, meeting the 90/90 accuracy/confidence criterion. The availability of finer spatial resolution (23.5 m) data of LISS-III of IRS-1C opened new vistas to obtain such information on crops growing in small holdings under multiple cropping situations. Studies were in progress for crop identification using microwave SAR data, with cloud penetration and all-weather capability. The multitemporal NOAA-AVHRR and IRS-1C/WiFS data could provide crop information at the regional level, useful for crop monitoring. A vegetation index was used as a surrogate parameter for assessing crop condition and yield estimation.

25. Information on soils, one of the most important basic natural resources, was needed for both agricultural and non-agricultural purposes. Remote sensing technology was highly useful for mapping soils at various levels. The factors affecting soil reflectance were highlighted and the methodology adopted for mapping soils using data from different sensors and land evaluation studies, including land capability, land irrigability and suitability for different crops, were illustrated by case studies. Watershed prioritization and the generation of optimum land-use plans were also discussed.

2. Presentations by country participants

26. The participants presented some of the GIS/remote sensing activities taking place in their respective countries with particular focus on sustainable rural development problems and issues.

Azerbaijan

27. The Institute of Ecology dealt mainly with the ecological monitoring of the environment through the development of software for thematic interpretation of remote sensing data and aerospace monitoring of the environment, including the Caspian Sea. Its activities included the definition and evaluation of the condition and dynamics of the water resources; the monitoring of water and soil pollution; the determination of the cloud cover condition; vegetation index monitoring; the inspection of anthropogenic and technogenic influences on the environment; and the study of the rapidly changing coast line of the Caspian Sea, its land and degradation. Other studies included yield forecasting for cotton and the development of maps showing the area of mud flows, droughts and landslide threats.

Bangladesh

28. The integration of remote sensing and GIS offered a potential means for rural development planning. Activities such as infrastructure development, agriculture applications such as crop monitoring, crop inventory, land budgeting, irrigation potentiality development and forestry could greatly benefit from such technologies. Bangladesh used remote sensing and GIS technology in its various management, planning and development activities and executed several projects in that field. The outcome of those studies provided valuable information for the development of rural Bangladesh.

Cambodia

29. The Integrated Resources Information Centre (IRIC), initiated in 1994 by the Cambodian Environmental Advising Team of UNDP and GIS/remote sensing office of the Ministry of Environment, would attempt to solve the main problems associated with the use of GIS and remote sensing in the country, namely the lack of up-to-date data and information, lack of knowledge to interpret and use such data, lack of training and human resource development in RS/GIS, lack of cooperation between agencies and duplication of efforts. An Environmental Information Centre of the MOE would be established in early 1997 and would act as the decentralized network for collaboration with and coordination among the public institutions, international agencies and NGOs.

China

30. Drought was the most important natural disaster for the Chinese agricultural production resulting in approximately 5 million tons of grain loss almost every year. An operational case study of winter wheat drought assessment and monitoring was undertaken in 1995 based on NOAA-AVHRR data in Huang-Huai-Hai region. The method of crop water stress index (CWSI = 1-E/Ep) and five drought classes (severe, intermediate, light, none and moist) was used based on the percentage of oil moisture content divided by soil field capacity of the upper 20 cm of soil. The results of both percentage of winter wheat water stress acreage divided by sowing acreage and soil moisture distribution maps for different provinces and counties per decade (between February and June) were described. Although comparison of the research results and field data showed 82 per cent of accuracy, comparative studies of different methodologies for drought assessment and monitoring were recommended.

Indonesia

31. Rural Community Development (RCD), a programme aiming to improve the quality of life of urban and rural communities, was initiated in 1957 by the Government by including it in its five-year development plan. Although much had already been achieved, further efforts were needed because of the large number of people (15 per cent) living under the poverty line. Remote sensing and GIS were considered useful tools for producing several types of maps essential for rural community development.

Islamic Republic of Iran

32. GIS was a very useful tool that could analyse large volumes of spatial data and produce information that met the user's needs. Soil erosion, a significant factor affecting national resources, was studied using GIS. Several models could be applied for erosion level assessment in a watershed basin where insufficient hydrologic data are available. The erosion potential model (EPM) was designed to predict erosion severity based on specific land-use/land-cover management systems. The data required for that model included land-use maps, slope maps and geological maps.

Malaysia

33. Malaysia had participated in several bilateral and regional projects in land use/cover mapping using satellite technology. The experiences gained from those projects had resulted in the operationalization of land use/cover mapping in the country on a large scale using optical satellite data. The major shortcoming encountered had been the occurrence of persistent and extensive cloud cover in certain areas, making it difficult to update land use/cover information on a regular basis. The use of SAR data, although still at an early stage, had met with some measurable success in alleviating that problem. The Malaysian scenario in land use/cover mapping using satellite data over the least 15 years was presented.

Mongolia

34. The development of integrated uses of remote sensing and GIS was proceeding at a relatively slow pace in the country due to the high cost of the technology. However, its potential was clearly recognized for land use/cover mapping, natural resource management, environmental monitoring and the production of sustainable rural development plans.

Myanmar

35. GIS applications were introduced in Myanmar with UNDP assistance through a series of forest inventory projects, a national forest inventory project and a pilot mangrove project. ICIMOD and other UNDP projects also assisted by providing some GIS hardware and software. While GIS facilities were found to be useful and helpful to the community, their high cost as well and that of Landsat data made the community approach essential. Extension services were also found to be vital, as communities were not usually environmentally conscious. The importance of training com-

petent staff to operate the GIS facilities was emphasized. It was felt that environmental and ecological degradation could be remedied through the participatory community approach with the assistance of remote sensing and GIS facilities. Participatory rural appraisals for development planning was described as a highly desirable and an inevitable choice.

Nepal

36. In Nepal, the ongoing eighth five-year plan (1992-97) had envisaged empowering the local governments of village, town and district levels by enhancing their capabilities in planning, implementation, monitoring and evaluation. Some programmes had been initiated to strengthen those local governments in database management. One example was the UNDP-supported project known as Participating District Development Programme (PDDP), implemented since 1993. At the National Planning Commission Secretariat, a GIS facility that helped local governments in collecting had been established to collect primary socio-economic and geographic information from the grass-root levels following a participatory approach, and helped in back-stopping the local governments through spatially-referenced databases.

37. Nepal needed information on natural resources for the rural development of the country since forests constituted one of the most important natural resources of the country. Remote sensing data had been providing supplemental data to the aerial photographs for forest mapping and monitoring. The National Remote Sensing Centre (NRSC) was established in 1981 as an autonomous institution through the joint cooperation between the Government of Nepal and USAID. The Centre had been working as a focal point for remote sensing activities concerning natural resources. NRSC was merged under the Forest Research and Survey Centre as the Remote Sensing Section of the Forest Survey Division in 1989. In the last few years, the main activities of the Remote Sensing Section consisted in collecting data on forest resources using conventional and modern techniques and providing data to all offices/agencies under the Ministry of Forests and Soil Conservation for sustainable management and development.

Republic of Korea

38. GIS was used to evaluate the cultivation area of red pepper in the country. The database developed for that purpose included weather data and soil maps. The range of values of various environmental factors affecting pepper growth were assigned scores based on their suitability for pepper. The crop suitability maps were produced through a simple summation of the scores.

Sri Lanka

39. The Mahaweli Authority of Sri Lanka (MASL) was responsible for the country's largest multi-purpose, trans-basin water resources development programme, based on the biggest river in the island Mahaweli. Under that programme, 52 per cent of the electricity, 22 per cent of rice, 50 per cent of onions, 70 per cent of chillies and many other crops were being produced. Large numbers of people (132,000 families) were being resettled. Under that programme, a large volume of data was being gathered and continued to be gathered, and it was realized that GIS provided the best tool for managing that data. To date, more than 5,000 sq km had been digitized at the 1:10,000 scale. MASL was currently planning to develop remote sensing capabilities, so that it could rapidly update the situation on the ground and monitor new developments and predict and guide the farmers as well as planners.

Thailand

40. A joint study project by the Land Development Department of Thailand and ITC of Netherlands was conducted with the study objective of GIS applications for the land use planning. The study examined some sub-districts of Petchabun Province in northern Thailand. The objective of the study was to produce a detailed up-to-date land-use map as a basis for land use planning. A soil map and multitemporal satellite images were used. A knowledge base was built up through extensive fieldwork. The knowledge base contained information on crop phenology and on the type of land

on which different agricultural crops were grown in the study area. Artificial intelligence techniques were used in a GIS environment to extract information on the current land use. A land-use map was produced showing the distribution of the main crops in the area. The resulting map, compared with the existing land-use map made from outdated aerial photographs, did not display very large discrepancies but showed improvements in areas where second crops were being cultivated. The localization of fields with maize was much more detailed than on the existing map. The method constituted an improvement over existing methods as it reduced the amount of field work necessary and improved the flexibility to combine with other sources of information.

Viet Nam

41. Principally an agricultural country, about 90 per cent of the population in Viet Nam were farmers. The rural environment was currently changing due to human activities such as deforestation, erosion, flood and environmental pollution. The routine updating of natural resource maps was seen as essential to improve the accuracy of prediction of future scenario. A case study of the Co To District was presented as an effort towards integrated remote sensing and GIS for rural sustainable development in Viet Nam.

3. Demonstration of application case studies

42. Application case studies in forestry, agriculture, geology, and water resources and in the management of land and water resources were demonstrated by NRSA experts to the workshop participants. The salient points are briefly described in the following paragraphs.

Forestry

(a) Temporal satellite data to monitor sustainable harvest of fuel and fodder needed by surrounding villagers in a forest reserve area;

(b) Temporal satellite data to detect illegal encroachments in forest reserve areas;

(c) Satellite data for broad category forest types mapping and inventory;

(d) Remote sensing and GIS for environmental impact study of a bauxite mining area in a forested environment;

(e) Remote sensing and GIS for identifying forest degraded zones meant for afforestation programmes with villagers' participation.

Agriculture

(a) Satellite data for crop acreage estimation via total enumeration for small areas and stratified random sampling approach in larger areas;

(b) Satellite data for reliable soil mapping in sparsely vegetated areas; mapping was made possible at family level;

(c) Remote sensing and GIS for the development of action plans for sustainable development. Basic thematic layers such as soil, land use/cover and degraded land zones were extracted from satellite data. In the GIS environment derived thematic layers such as land suitability classification for crops and land irrigability classification were generated in the GIS. The development of action plans made use of those derived layers and other composite layers such as erosion risk zones;

(d) Monitoring of shifting cultivation activity in a forest environment could be done using change detection techniques and satellite data.

Geology

(a) Satellite data for extraction of thematic information -- rock outcrops, land-forms, lineament patterns and forest density. GIS was used to output composite suitability zone maps for the location of a hydroelectric project based on those information layers;

(b) Through specialized enhancement techniques employed in satellite data, small bauxite mining areas could be located and their environmental impacts assessed in the GIS;

(c) The utilization of Landsat TM band 6 to detect temporal temperature difference, a useful precursor indicator for impending earthquakes.

Water resources

(a) Remote sensing and GIS to assess irrigation status in Thunga Bhadra area. Satellite imagery was used to estimate sedimentation in the reservoir, and the water area extent in turn was used to estimate the corresponding water storage capacity of the reservoir via correlation graphs plotted by hydrographic survey of the dam.

Geographically Encoded Land and Water Information System (Geo-LAWNS)

(a) Geo-LAWNS (Geographically Encoded Land and Water Information System) was essentially an integrated remote sensing and GIS package that provided value-added service for land and water resource development planning and management on a sustainable basis;

(b) NRSA was also developing a software package for customizing crop acreage and production estimation involving digital image processing procedures for sample segmentation and boundary masking on heterogeneous platforms.

4. Field visit

43. A trip to the Earth Observation Satellite Receiving Station was carried out on the final day of the workshop. A demonstration in the field of the grass-roots implementation of IMSD also took place in the village of Mahaboobnagar.

5. Conclusions and recommendations

44. The Workshop emphasized that the integrated use of remote sensing and GIS was important, proven and efficient for addressing information needs for sustainable rural development. It strongly recommended that such technology applications should be integrated with national development activities.

45. The Workshop acknowledged with appreciation the promotional efforts of ESCAP and UNDP in the region through the Integrated Uses of Remote Sensing and GIS for Sustainable Development Programme. Such efforts, along with the activities initiated by several countries, had greatly contributed in getting the regional countries to increasingly benefit from the technology applications to sustainable rural development.

46. The Workshop noted that the most important factors related to rural development needed to be identified so that they could be prioritized and considered according to budgetary constraints. It was agreed that remote sensing and GIS technologies should be integrated wherever possible in any development activity and that development activities should be adequately monitored using such technologies in order to identify any adverse impacts and take the necessary corrective measures so that a balance could be obtained between development, on the one hand, and ecology and environment on the other.

47. Sustainable rural development was recognized as a very important issue and the Workshop agreed that the sharing of experience in that field between different countries of the region through coordinated events such as the current workshop should be continued on a more frequent basis.

48. The Workshop believed that coordination was needed at different levels and touched many aspects such as standardization, human resource development, infrastructure development and improvement of the interface between the grass-roots users and the GIS/remote sensing specialists. One way to address such coordination issues at the national level could be through the creation of national coordinating bodies or agencies that would interface with the various user agencies and

thus help in avoiding duplication of efforts and in enhancing interfacing and coordination, particularly with respect to rural development.

49. The Workshop observed that the sharing of data and information between the different government agencies in a single country was still a major hurdle to the development of useful national databases accessible to the potential users. Further efforts needed to be expended in educating the decision makers about the needs for data sharing for the benefit of the country.

50. The issue of standardization in terms of thematic classification, data format, output product specification and so on was recognized as critical to facilitate the wider use and sharing of the databases constructed by the concerned agencies, and absolutely necessary for multi-disciplinary activities such as sustainable rural development planning. Therefore, it was considered that that issue should be given close attention at both national and regional levels.

51. Interface with farmers, the ultimate users of the technology in the rural development context, was seen as requiring special attention to find innovative ways to effectively communicate the information provided by GIS/remote sensing technology and the advantages it offers over the methods traditionally used. The involvement of the private sector and NGOs might be considered to further publicize successful rural development initiatives.

52. The development of national databases using remote sensing and GIS was seen as extremely important and requiring coordination between the various sectoral agencies. Those, in turn, needed to closely interact with people at the grass-roots level to involve them in the data collection and development planning process. The Workshop emphasized that community participation throughout all phases of a development project, including the decision-making process based on information provided through GIS/remote sensing technologies, was essential in order to increase the probability of success of such projects. In that connection, appropriate site-specific decision-making models needed to be developed.

53. The knowledge and skills required for operating various remote sensing/GIS systems were an important component and hence the Workshop recognized the need of user-friendly facilities and tools, and the customized training of technical staff. Those should be considered as an important project component so that the technology could be widely used for various fields of planning and decision-making.

54. The Workshop agreed that many developing countries of the region had embarked on GIS/ remote sensing projects at different levels but to a rather limited extent. It was firmly believed that strong government involvement and backing were essential to ensure widespread use of that technology and its applications at the grass-roots level. The successful experiences observed in countries such as India needed to be further publicized to developing countries of the region to stimulate greater government backing.

55. The acceptance of the remote sensing/GIS technology in rural development very much depended on the understanding of planners and decision makers about the merits and applications of the technology. Therefore, promotional efforts to expose high-level planners and decision makers should be organized at national and regional levels.

56. The Workshop strongly requested that the ESCAP secretariat enhance its activities in the region to assist the member countries in building national capabilities in the use of remote sensing and GIS for sustainable rural development. It also strongly recommended that UNDP continue its partnership with the ESCAP secretariat in promoting wider remote sensing and GIS applications in its future programmes. It also strongly recommended that ESCAP develop its partnership with regional organizations working in that field to enhance the delivery of services to member countries and in human resource development.

57. The Workshop recommended that ESCAP work in close collaboration with the Department of Space of India to help in capacity-building and make IRS data available to developing countries of the region.

58. The Workshop observed that vast quantities of satellite remotely sensed data were needed to carry out the suitable mapping and monitoring of natural resources and the environment in any given country. For many countries, the cost of such data constituted a major barrier to the construction of the databases required for proper sustainable development, including rural development. It was felt that developed countries should increase their investments to developing countries to improve the accessibility to satellite data and spatial information system technology.

59. The Workshop agreed that greater regional cooperation should be encouraged to solve trans-boundary issues and problems. Pilot projects between countries sharing similar problems were seen as a desirable way to encourage such cooperation. ESCAP might continue to coordinate such activities in the region.

6. Adoption of the report

60. The report of the workshop was reviewed and adopted on 21 September at 16:45. Before the workshop was closed, the participants and the representatives of the ESCAP secretariat once again expressed their deep gratitude to the Government of India for their warm hospitality, and particularly to the National Remote Sensing Agency for its excellent preparatory work and arrangement under the guidance of the Department of Space.

PART TWO

OVERVIEW OF REMOTE SENSING AND GEOGRAPHIC INFORMATION SYSTEMS

INTEGRATED REMOTE SENSING FOR RURAL DEVELOPMENT IN INDIA: AN OVERVIEW

*D.P. Rao**

A. Introduction

The Indian space programme has made great strides in using sophisticated remote sensing technology to bring maximum benefit to the people at the grassroots level of the nation in a timely and cost-effective manner. India is one of the few developing countries to have its own indigenous operational space and ground segment as well as necessary expertise and infrastructure for the operational use of remote sensing techniques for development projects. India is a large country endowed with many natural resources. These resources are spread over a vast stretch of land, with an area of 329 million hectares, and a much larger ocean area coming under the Exclusive Economic Zone. There is increasing pressure on the resources of land and ocean with India's population crossing the 900 million mark. Pressure will mount further as the population is expected to touch the 1,100 million mark in the next two decades. Optimum utilization of land resources and maximizing production hold the keys to the problem. In India, remote sensing technology plays a pivotal role by providing means for keeping a close watch on conservation, monitoring and management of natural resources.

The green revolution in India enabled the country to increase its food production from 55 million tons in 1947 to 177 mt in 1990, and the target of reaching 220 mt to meet the basic needs of the projected one billion population by the year 2000 will be a formidable one, particularly considering that 47 per cent of the nation's land area is already under agriculture. The average yield of rice in the country is just about 1.7 tons per hectare, as against 5.0 tons/ha in the United States. The average yield of wheat in India is about 2.2 tons/ha as against 3.7 tons/ha for the world average and 5.4 tons/ha in the United States. It is worth mentioning here that the "green revolution" and increased productivity in India and other developing nations is mainly due to increased irrigation of over 30 per cent of the land area. Large-scale irrigation, extensive use of chemical fertilizers, inadequate drainage and bad agricultural practices have resulted in soil salinity, making some of the most fertile land unproductive and degraded. Almost 25 per cent of the arable land in every continent has become problem land, with another 25 per cent having very low productivity. It can be cited here as an example that the fertile Indo-Gangetic plain in Uttar Pradesh and Bihar, which was once the cradle of civilization, now accounts for only 50 per cent of the sugarcane growing area in India, producing just 40 tons/ha, because of the high level of salinity, as against achieved yield of 90-100 tons/ha in Karnataka, Maharashtra and Tamil Nadu. Highly fragmented land holdings, low-input agricultural practices and mismanagement of water resources have further added to the problems faced by India and the world's developing countries in their effort to improve their agricultural output. It has been said that even today over 600 million people in the world and 200 million people in Asia alone are underfed and undernourished, so it is clear that, without doubling the present rate of growth in agricultural production to 7 per cent per year, the coming decades will witness the greatest tragedy of humankind in history -- that of millions dying from starvation. The growing pressure of exploding population, increased demand for food, fodder and fuelwood, combined with intense industrial activity have essentially led to rapid deforestation in India as well as in the world as a whole. In India the closed forest has been depleted from 14 per cent to 11 per cent of the land area in over a period of one decade. Such severe deforestation has not only been the cause of climatic changes but also has led to soil erosion (almost 10 tons/ha per year as opposed to 1 ton/ha in the forested area), resulting in sedimentation of reservoirs and rivers and causing recurrent floods and increasing desertification. Whereas in the past it was climatic fluctuations that

* National Remote Sensing Agency, Hyderabad, India.

were responsible for the expansion and creation of deserts, today the main factors responsible for increasing desertification are overgrazing, deforestation and general mismanagement of land and water resources. About 3,000 million ha, a quarter of the Earth's land surface, is now turning into desert or is being damaged by factors that contribute to desertification.

B. Water resource management

Since India is an agriculture-based economy, management of water resources is given top priority, and remote sensing data are routinely used in the country for water resource management, such as prioritization of watersheds, monitoring of surface water bodies, rainfall run-off studies and irrigation scheduling. Using satellite data in conjunction with collateral information on rainfall, soil types and land-use patterns, it has been possible to estimate erosion rates, plan erosion control measures and identify sites for water harvesting. Satellite remote sensing has become an invaluable tool in India for predicting snow-melt run-off from the Himalayan snow cover river basins for optimal planning of water resource use for power and irrigation in north India. For this purpose, not only snow cover information, but use of digital elevation models, snow depth and temperature is being attempted to improve the snow-melt run-off prediction.

In India, a country which receives a bountiful amount of annual rain (about 130 cm/year), the situation of its water resources, both surface water and groundwater, has assumed alarming proportions -- the underground water table in many areas of the country is depleting at a rate of about 10 metres/year, because of the lack of any viable recharging mechanism. Optimal management of water becomes crucial particularly in the dryland tracts of the country where most of the rainfall occurs on fewer than 60 days. With the added problems of higher temperature regimes and higher evapotranspiration rates in these critical areas, the need for optimal harvesting of run-off and the recharging of underground aquifers assumes paramount importance. Although major irrigation projects and big dams have contributed to improved agricultural production in the last few decades, the problem of waterlogging, salinization and loss of valuable bio-resources have led to the gradual degradation of land in many areas in the developing world including India. On the other hand, it is well recognized that small ponds and check dams can improve the efficiency of water harvesting and soil moisture conservation without affecting the ecological system. Intensive use of chemical fertilizers and pesticides combined with poor management of watersheds has resulted in severe water stress, pesticide contamination in the soil as well as the water, and the degradation of soil resulting in the disruption of ecosystems over large areas. Acid precipitation has not only affected aquatic life and fish shoals but also soils in some areas where acidification has affected the soil to a depth of almost one metre.

C. Land resource management

The phenomenal population growth from 350 million to 900 million in the 45 years since independence in 1947 and the sub-optimal management of land and water resources are eroding the substantial gains that have been achieved from the "green revolution" and "operation flood". The limited land available for agricultural production, about 0.17 ha per person, is already becoming degraded because of poor agricultural practices. About 15-20 per cent of India's land area is considered as wasteland because of salinity of soil caused by excessive use of fertilizers, improper irrigation practices, degradation due to prolonged agricultural usage, use of slash and burn clearing techniques, erosion of fertile soil due to overland flow of water and due to the spread of the desert. Under the aegis of the National Wasteland Development Board, a project was taken up for wasteland mapping of 146 critically affected districts of India using 1:50,000-scale satellite images. The results of this mapping helped identify thirteen recognizable wasteland categories covering about 23 million ha, almost half of which, with some interventions, can be reclaimed for agricultural production. Actions have been initiated by concerned developmental sectors of the country to reclaim these culturable wastelands by afforestation, fodder development and other agricultural activities like oil-seed production and horticulture.

D. Conceptual framework of Integrated Mission for Sustainable Development

As these examples illustrate, the world is not now headed toward a sustainable future, but rather toward a variety of potential human and environmental disasters. Over the past 20 years, the world has started recognizing that environmental problems are inseparable from those of human welfare and from the processes of economic development. So there is an imperative need to look again into the various factors to be considered towards a change -- for sustainable development. Development of any kind, whether social, economic, educational or technological, is a process of "change". Some changes, such as giving up some of the older practices or older methods and adoption of newer practices and newer methods, are required for development to take place. Changes could be for the better or could be for the worse. It is indeed when we want development to take place that we want changes to take place for the better. If abrupt changes take place, the process could be quite disruptive and the resultant changes could lead to irreversible damages. Therefore, any change process should take place in such a manner that minimum disruption is caused; hence is it necessary that the changes should occur gradually. The changes should also occur with fully informed understanding of the consequences of the changes and conscious participation of the people, involved both in the planning as well as execution process, if these changes are to result in sustainable activities.

Therefore, a new developmental path is required, which is one that will sustain human progress not just in a few places for few years, but for the entire planet into the distant future. Sustainable development is defined as development that meets the needs of the present without compromising the ability of future generations to meet their own needs. Sustainable development can also be defined as development of natural resources not only to meet the immediate needs of the present population but also to meet the requirements of future generations without in any way endangering either their productive potential, the environment, or ecological systems. In other words, we need to take into account the local as well as the global effects to arrive at an optimal solution for meeting the basic aspirations of the affected people.

The concept of sustainable development has been widely accepted and endorsed. However, translating this concept into practical goals, programmes and policies upon which nations could agree will be a hard job to accomplish and impart because nations face widely varying circumstances.

Primary production activities aimed at higher levels of production than the optimum potential of the land meet with frequent failures leading to low production and land degradation in the long run. Therefore, there is an urgent need to reverse such trends by bringing the land back under appropriate land-use practices. The Integrated Land and Water Management Study for Sustainable Development aims at increasing production from and productivity of natural resources, especially land and water, on a sustainable basis while maintaining an ecological balance. Sustainable development of natural resources relies on maintaining the fragile balance between productivity functions and conservation practices through precise identification and systematic monitoring of problem areas in various resources and developmental sectors. Moreover, it calls for application of alternate agricultural practices, crop rotation, use of bio-fertilizers, energy-efficient farming methods and reclamation of under-utilized land and wastelands, planned exploitation of mineral and groundwater resources, among other measures. Optimal exploitation of the resources (both renewable and non-renewable) with proper enriching mechanisms calls for cutting across the narrow confines of sectoral approaches, taking a holistic view of the region. As a first step, using space technology, natural resources such as soil, groundwater, surface water and natural vegetation are studied and terrain analysis is made for land-forms and land use, in addition to studies on slope, watershed and drainage. Climatic factors like rainfall, temperature, humidity and wind speed are also collected and analysed. All the above parameters are integrated to evaluate the production potential of the land, keeping in view the advances in technological development for primary production activities such as agriculture, horticulture, fodder and pasture development, timber production and pisciculture. A flow chart showing various steps and activities of the integrated study for sustainable development in the context of district-level planning is given in figure 1. Integrated farming systems -- such as food crops-horticulture-fodder/pasture, or food crops-vegetables-pisciculture-fodder/pasture, or horticulture-vegetables-forest plantation-fodder/pasture -- with efficient irrigation and water management practices and soil and water conservation measures are emerging as appropriate land-use practices for sustainable production and productivity. Geographic information systems (GIS) have been used for integration of spatial data on various resource themes. Suitability of various combinations of land parameters, such as soil, ground water quality and potential, slope, land form, land use and land cover, have been linked to primary production activities through the rule-based decision capabilities of GIS packages. On the basis of the studies, alternate development plans, showing site-specific primary production activities, are prepared. Remote sensing technology has played a key role in inventorying and mapping natural resources, land use, land cover and land forms.

E. Remote sensing applications

With the advent of remote sensing in the early 1970s, after the launching of the Landsat series of satellites by NASA, the SPOT satellite by France and the Indian Remote Sensing (IRS) series of satellites by the Indian Space Research Organization (ISRO), Department of Space (DOS), there has been an increasing utilization of satellite imagery for inventory and monitoring of natural resources in India and abroad. Remote sensing techniques have proved to be rapid and cost-effective and are able to provide timely information with their repetitive coverage of the same area within a span of 5 to 22 days. Realizing the capabilities of IRS-1A, 1B and now 1C data, the DOS initiated a pilot project named Integrated Study to Combat Drought on a sustainable basis in 21 perennially drought-affected districts of India. The objective of this study was to evolve action plan packages for management of surface and groundwater, agriculture and fodder resources by integrating thematic information on natural resources and socio-economic information.

These pilot studies were carried out by DOS along with remote sensing application centres of concerned states, with active involvement of district officials. Apart from the above organization, agricultural universities, research institutions and voluntary agencies were also involved in this task.

The resource maps such as soil, land-use and hydro-geomorphology, have been prepared using Indian Remote Sensing satellite images at 1:50,000 scale. The physical features like drainage, slopes, watershed, road and railways with settlement location and village boundaries are also mapped on the same scale. This natural resource information is integrated with socio-economic and demographic data in order to prescribe a suitable site-specific action plan for appropriate land-use, fodder and water management practices. The results of the pilot studies for six districts out of 21, namely Anantpur (Andhra Pradesh), Ahmednagar (Maharashtra), Bhiwani (Haryana), Dharmapuri (Tamil Nadu), Jhabua (Madhya Pradesh) and Kalahandi (Orissa), were presented to the Deputy Chairman and members of the Planning Commission and secretaries of central and state departments. Appreciating the methodology developed by DOS, the Planning Commission desired to take up six small watersheds of about 10,000 hectares each from the above-mentioned districts for implementation, in a phased manner, of the action plan that has emerged from this study.

Subsequently, a high-level committee set up by the Planning Commission has recommended to take up similar studies in 153 districts that are affected by a variety of problems such as drought, wasteland, hilly terrain and flooding. Even though entire districts are problem areas, initially the study will be confined to some priority *taluks*/blocks/watersheds in each district. The study will be jointly financed by DOS and respective state governments. This major national project has been named the Integrated Mission for Sustainable Development (IMSD) and was launched in June 1992 by DOS.

Analysis of remote sensing and conventional database information leading to locale-specific action plans will be generated by the respective state remote sensing centres with the technical guidance of various units of DOS. The developmental/technological inputs will be obtained from line departments of central and state governments, agricultural universities and research institutions.

Implementation of the action plans will be undertaken by district authorities under the guidance of the state governments with the active involvement of voluntary agencies. Expert committees constituted specifically for each state and district will review the action plans before they are taken up for implementation. The Mission Director from DOS and project directors for each state, as well as from DOS, have been identified for execution of the study. Close linkages with the central agencies -- the Department of Science and Technology (Natural Resources Management System), National Informatic Centre, District Information System, Central Groundwater Board, Central Water Commission, National Water Development Board, Forest Survey of India, Indian Council of Agricultural Research, National Bureau of Soil Survey and Land Use Planning, All India Soil and Land Use Survey, among others -- will be established. In addition, numerous voluntary agencies with expertise in resource management at the grassroots level will be involved during the implementation stage. The scheme of execution for IMSD is given in figure 2.

As an example, brief highlights of the action plan prepared for Vanjuvanka watershed in Anantpur District, Andhra Pradesh, are described below:

F. Case study

1. Vanjuvanka watershed in Ananthapur District, Andhra Pradesh

In Andhra Pradesh, the Integrated Study to Combat Drought on a Sustainable Basis was taken up in Ananthapur District. A watershed covering 13,452 ha, falling in parts of eight revenue villages, was selected for the study. The watershed receives an average rainfall of 300 mm annually in 23 rainy days. The area represents a semi-arid climate with undulating topography and is hence subject to quick surface run-off and erosion hazards. The vegetation cover is meagre and there are no surface water sources except seven minor irrigation tanks. The soils are shallow and poor in nutrient status. The groundwater table is very deep due to over-exploitation resulting from inefficient water management practices. There is an urgent need to augment the fodder resources in the project area, in view of a large livestock population.

The thematic information, i.e. hydro-geomorphology, land use and land cover, soils, drainage networks and other collateral data derived from Survey of India toposheets (1:50,000 scale), along with available socio-economic data, were spatially integrated and evaluated in the context of available current technology, leading to the generation of action plans for land and water resource development in the watershed area. Some of the suggested activities for water resource development in the watershed area include construction of rainwater harvesting structures such as percolation tanks, check dams, farm ponds, subsurface dykes, diversion drains and other structures.

For land resource development, various optimal land use and farming systems were recommended, including changes in cropping pattern (for example, inclusion of vegetable crops, mulberry and other plants at suitable locations), soil and moisture conservation measures (contour bunding and vegetative barriers), alternative land-use systems (horticulture, agroforestry, fodder and fuelwood plantation and so on) and afforestation measures in degraded forest areas.

The action plan map was jointly prepared by scientists from the National Remote Sensing Agency (NRSA) and Andhra Pradesh Remote Sensing Application Centre, Hyderabad, in consultation with scientists from APAU (Regional Research Station, Ananthapur), district administrative officials, local people and NGOs working in the area.

The action plan was reviewed by an expert committee consisting of scientists drawn from various central, state and international research institutes and officials of various government departments for its implementation. The District Collector, Ananthapur, was identified as the focal point for implementation of the action plan by involving local NGOs for facilitating mass participation by people.

2. Achievements in the watershed area

The implementation of prescriptions made under this project has resulted in the construction of several check dams and contour bunding, besides other development activities. Due to construction of check dams, there was a recharge of groundwater in downstream areas leading to 1-4 metre rises in the water table in dug wells.

In view of the highly encouraging results of IMSD implementation, the Ministry of Rural Areas and Employment has requested DOS to take up a special IMSD study for 92 blocks in 92 DPAP districts of the country.

G. GIS applications

A geographic information system is essentially a tool that helps a planner in decision-making. As a part of the National Resources Information System (NRIS) programme, several pilot studies on the use of GIS have been carried out by DOS. These are urban planning, action plan development for district planning under IMSD, wasteland development, watershed prioritization, mineral targeting and flood management, to name a few important areas of application. The users are being trained not only in handling remote sensing data but also in maintaining the database in GIS for decision-making and updating of the database.

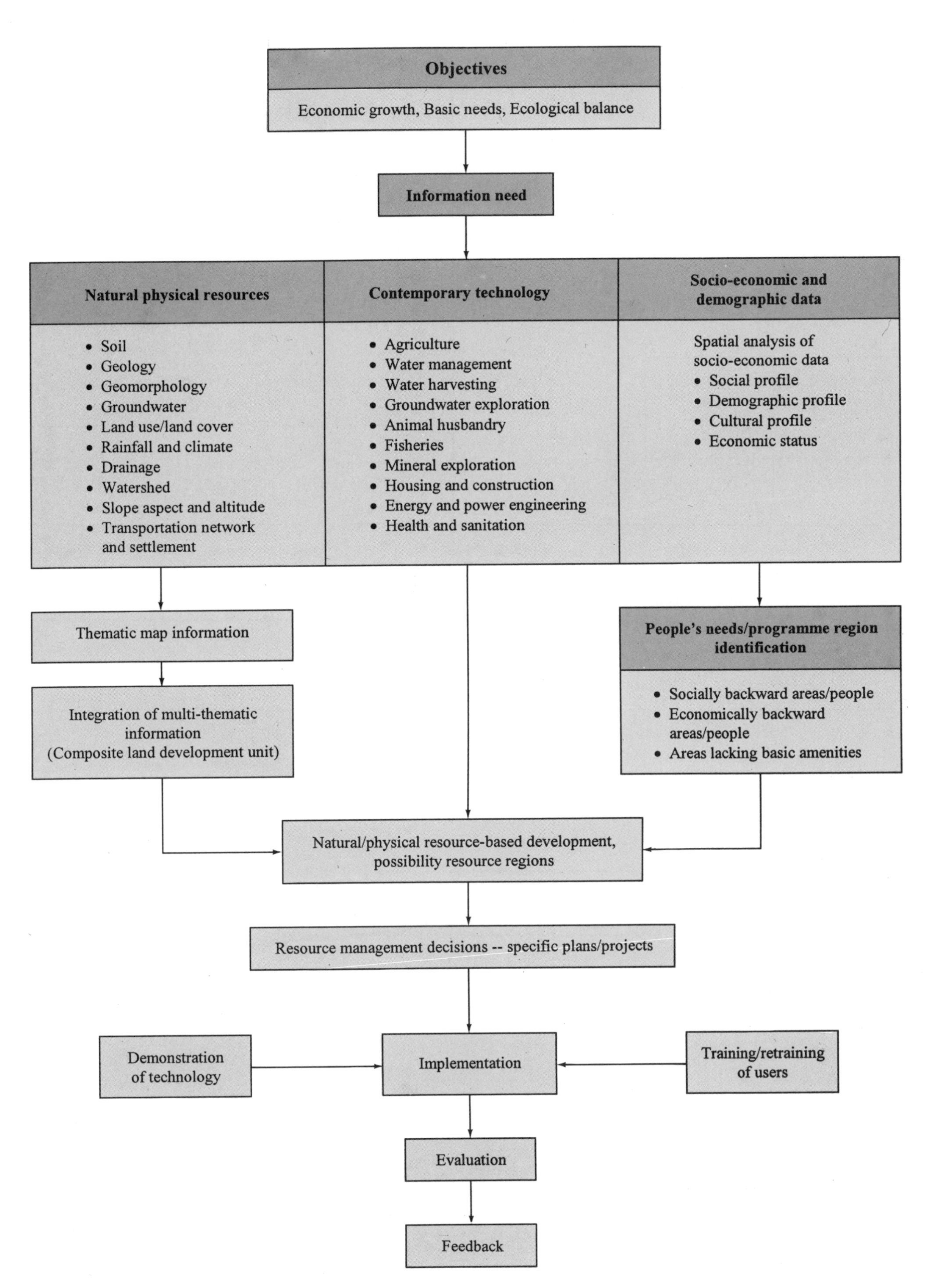

Figure 1. Conceptual framework of IMSD

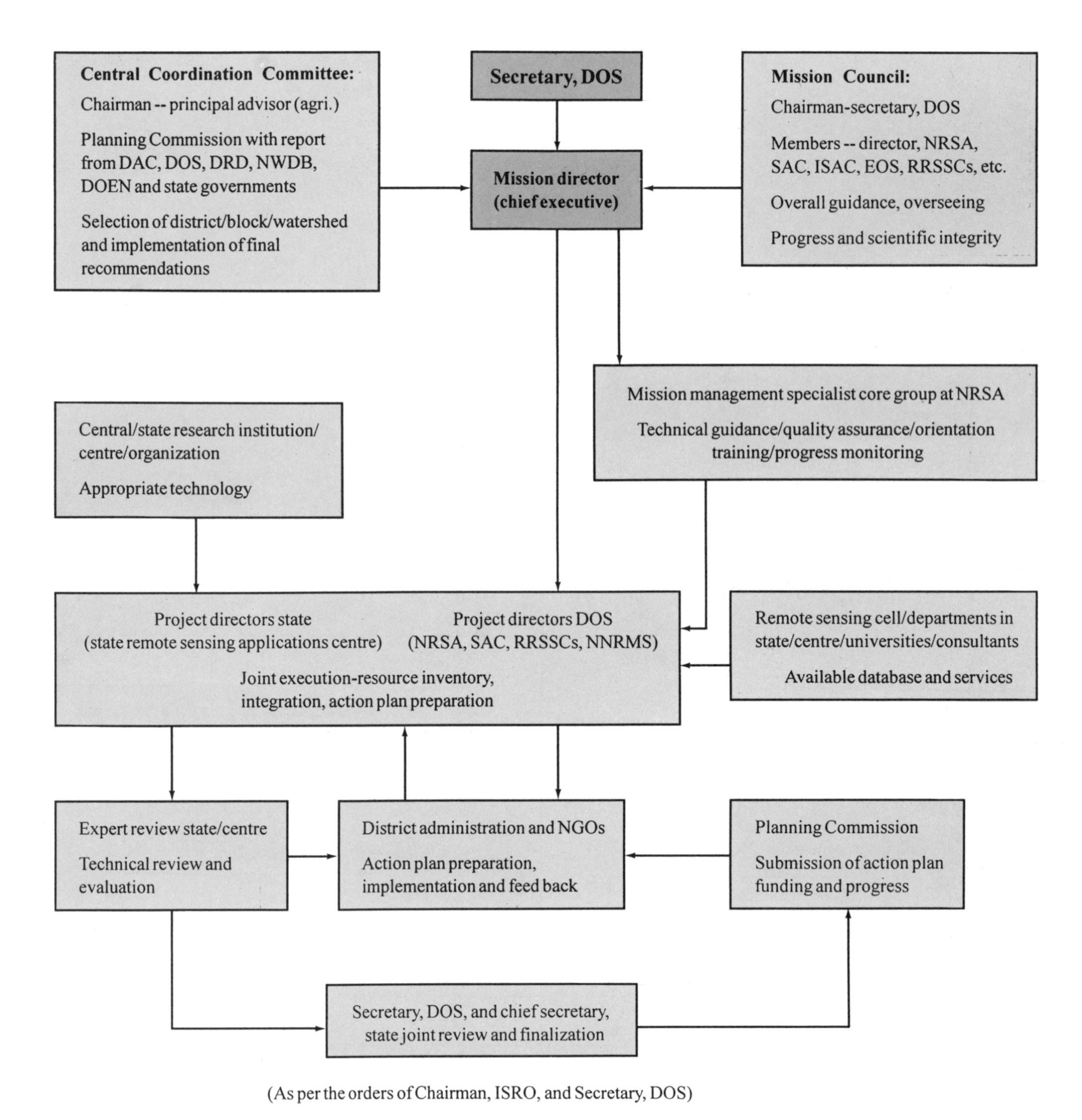

Figure 2. Scheme of execution of IMSD

GEOGRAPHIC INFORMATION SYSTEMS: AN OVERVIEW

*T. Ch. Malleswara Rao**

A. Introduction

Natural resources are great assets for any country's economy. Large reserves of resources are still unexplored and underutilized. Sectorally oriented and independent exploitation creates serious ecological problems. An integrated approach to the exploitation, conservation and management of natural resources in their totality, as well as to the protection and preservation of the environment, is essential. The exploration, exploitation and judicious management of natural resources require a sound system for the development of a region. In some of the above functions, integrated geographic information systems (GIS) and image analysis technologies play an important role in developmental planning, implementation and monitoring.

B. Geographic information systems

The collection of data about the spatial distribution of significant properties of the Earth's surface has long been an important part of the activities of organized societies. From the earliest civilizations to modern times, spatial data have been collected by navigators, geographers and surveyors and rendered into pictorial form by the map makers or cartographers. In the eighteenth century, European civilization once again reached a state of organization such that many governments realized the value of systematic mapping of their lands. During the last 200 years many individual types of maps have been developed, but there is a long, unbroken tradition of high cartographic standards that has continued until the present.

As scientific study of the Earth advanced, so new material needed to be mapped. The developments in the assessment and understanding of natural resources -- geology, geomorphology, soil science, ecology and land -- which began in the nineteenth century and have continued to this day, provided new material to be mapped. Whereas topographical maps can be regarded as general purpose because they do not set out to fulfil any specific aim (i.e. they can be interpreted for many different purposes), maps of distribution of rock types, soil series or land use are made for more limited purposes. These specific-purpose maps are often referred to as "thematic" maps because they contain information about a single subject or theme.

The theme may be qualitative (as in the case of land-use classes) or quantitative (as in the case of the variation of the depth to the phreatic zone). Both quantitative and qualitative information can be expressed as a choropleth map, that is, areas of equal value separated by boundaries; typical examples are soil maps, land-use maps or maps showing the results of censuses (figure 1). Quantitative data can also be mapped by assuming that the data can be modelled by a continuous surface that is capable of mathematical description. The variations are then shown by iso-lines or contours. Typical examples are the elevation contours on a topographic map and the isobars of a weather chart (figure 2).

The collection and compilation of data and the publication of a map is a costly and time-consuming business. Consequently, the extraction of single themes from a general purpose map can be prohibitively expensive if the map must be redrawn by hand. It was not important that initial mapping costs were large when a map could be thought of as being relevant for a period of 20 years or more. But there is now such a need for information about how the Earth's surface is changing that conventional map-making techniques are totally inadequate. There has been much duplication of effort and a multiplication of discipline-specific jargon for different applications in different lands. This multiplicity of effort in several initially separate but closely related fields is now resulting in

* National Remote Sensing Agency, Hyderabad, India.

the possibility of linking many kinds of spatial processing together into truly general-purpose geographic information systems, as technical and conceptual problems are overcome (figure 3).

Essentially, the hand-drawn map or the map in a resource inventory is a snapshot of the situation seen through the particular filter of a given surveyor in a given discipline at a certain moment in time. More recently, the aerial photograph, but more especially the satellite image, has made it possible to see how landscapes change over time, to follow the slow march of desertification or erosion or the swifter progress of forest fires, floods, locust swarms or weather systems. But the products of airborne and space sensors are not maps, in the original meaning of the word, but photographic images or streams of data on magnetic tapes. The digital data are not in the familiar form of points, lines and areas representing the already recognized and classified features of the earth's surface, but are coded in picture elements -- pixels -- cells in a two-dimensional matrix that contain merely a number indicating the strength of reflected electromagnetic radiation in a given band. New tools were needed to turn these streams of numbers into pictures and to identify meaningful patterns. Cartographers, initially, did not possess the skills to use these new tools and so the fledgling sciences of remote sensing, image analysis and pattern recognition were nursed into being, not by the traditional custodians of spatial data, but by mathematicians, physicists, computer scientists and military authorities. These new practitioners of the art of making images of the Earth have taken a very different approach to that of the conventional field scientists, surveyors and cartographers. In the beginning, they often made exaggerated claims about the ability of remote sensing and image analysis to recognize and map the properties of the Earth's surface without expensive ground surveys. Gradually it has come to the realized that the often very striking images produced from remotely sensed data only have a real value if they are linked to ground truth; therefore, a certain amount of field surveying is essential for proper interpretation. And to facilitate calibration, the images have to be located properly with respect to a proper geodetic grid, or otherwise the information cannot be related to a definite place. The need for a marriage between remote sensing, Earthbound surveys and cartography arose. And it has been made possible by the class of mapping tools known as geographic information systems, or GIS.

1. The components of a geographic information system

Geographic information systems have three important components: computer hardware, sets of application software modules, and a proper organizational context. These three components need to be in balance if the system is to function satisfactorily.

(a) Computer hardware

The general hardware components of a geographic information system are presented in figure 4. The computer or central processing unit (CPU) is linked to a disk drive storage unit, which provides space for storing data and programs. A digitizer or other device is used to convert data from maps and documents into digital form and send them to the computer. A plotter or other kind of display device is used to present the results of the data processing, and a tape drive is used for storing data or programs on magnetic tape, or for communicating with other systems. Inter-computer communication can also take place via a networking system over special data lines, or over telephone lines using a device known as a "modem". The user controls the computer and the peripherals (a general term for plotters, printers, digitizers and other apparatus linked to the computer) via a visual display unit (VDU), otherwise known as a terminal. The user's terminal might itself be a microcomputer, or it might incorporate special hardware to allow maps to be displayed quickly.

(b) GIS software modules

The software package for a geographic information system consists of five basic technical modules (figure 5). These basic modules are sub-systems for:

(a) Data input and verification;
(b) Data storage and database management;

(c) Data output and presentation;
(d) Data transformation;
(e) Interaction with the user.

Data input (figure 6) covers all aspects of transforming data captured in the form of existing maps, field observations and sensors (including aerial photography, satellites and recording instruments) into a compatible digital form. A wide range of computer tools is available for this purpose, including the interactive terminal or VDU, the digitizer, lists of data in text files, scanners (in satellites or aeroplanes for direct recording of data or for converting maps and photographic images) and the devices necessary for recording data already written, i.e. magnetic media such as tapes, drums and disks. Data input and the verification of data are both needed to build a geographic database.

Data storage and database management (figure 7) concerns the way in which the data about the position, linkages (topology) and attributes of geographical elements (points, lines and areas representing objects on the Earth's surface) are structured and organized both with respect to the way they must be handled in the computer and how they are perceived by the users of the system. The computer programme used to organize the database is known as a database management system (DBMS). Data output and presentation (figure 8) concern the ways the data are displayed and how the results of analysis are reported to the users. Data may be presented as maps, tables and figures (graphs and charts) in a variety of ways, ranging from the ephemeral image on a cathode ray tube through hard-copy output drawn on printer or plotter to information recorded on magnetic media in digital form.

Data transformation embraces two classes of operation, namely (a) transformations needed to remove errors from the data, bring them up to date or match them to other data sets, and (b) the large array of analysis methods that can be applied to the data in order to achieve answers to the questions asked of the GIS. Transformations can operate on the spatial and the non-spatial aspects of the data, either separately or in combination. Many of these transformations, such as those associated with scale-changing, fitting data to new projects, logical retrieval of data, and calculation of areas and perimeters, are of such a general nature that one should expect to find them in every kind of GIS in one form or another. Other kinds of manipulation may be extremely application-specific, and their incorporation into a particular GIS may be only to satisfy the particular users of that system.

(c) The organizational aspects of GIS

The five technical subsystems of GIS govern the way in which geographic information can be processed but they do not of themselves guarantee that any particular GIS will be used effectively. In order to be used effectively, the GIS needs to be placed in an appropriate organizational context. Just as in all organizations dealing with complex products, like the manufacturing industry, new tools can only be used effectively if they are properly integrated into the whole work process and not tacked on as an afterthought. To do this properly requires not only the necessary investments in hardware and software, but also in the retraining of personnel and managers to use the new technology in the proper organizational context. In the 1970s the high price of many commercial systems sold for geographic information processing made managers cautious of making expensive investments in the then new and untried technology. In recent years the falling hardware prices have encouraged automation, but skilled personnel and good, reasonably priced software have remained scarce. There are still many choices open to an organization wishing to invest in geographic information systems.

C. GIS and other technologies

The development of the computer-based technology for spatial data handling has both drawn upon and contributed to a number of other technical areas. The main conceptual development of these systems has come, of course, from geography and cartography, but their present effective status would not have been possible without a number of significant, interdisciplinary interactions.

1. Computer graphics/image processing

The output, or reporting, state of the geographic information system is heavily dependent upon the availability of rapid, high resolution, graphics displays. We have been fortunate in that these displays have also been of significant utility in a substantial number of other fields. This high demand level has led to the rapid development of low-cost, sophisticated, computer graphics systems that are now capable of reproducing any desired spatial data display. Indeed, the development of this technology (especially in the area of colour rendition and dynamic displays) has now outstripped our effective ability to make use of it.

Computer graphics, especially image processing, has cost-effective and sophisticated hardware. Many of the algorithms used in computer graphics and the data structures used in image processing have proven quite useful in spatial data handling. Conversely, a number of developments pertaining to algorithms and data structures for spatial data handling have proven to be of considerable utility in the computer graphics area.

2. Computational geometry

A specialized, and hence rather small, area of computer science deals with the analysis of algorithms for handling geometric entities. The work that has been undertaken here has led to significant improvements in geographic information systems (e.g. the recent development of the Arc Info system by ESRI, Inc.), and has stimulated a growing interest in the explicit analysis of the efficiency of algorithms used in spatial data handling systems.

Although the number of persons involved in this area of computer science is small, their work has had a disproportionate impact on GIS development.

3. Database management systems

In contrast to computational geometry, theoretical and practical work on systems for managing large volumes of data has occupied the attention of a substantial number of academic and commercial researchers in computer science. Although a number of these systems have been applied to simple forms of spatial data (e.g. point data), their developmental emphasis on one-dimensional data has limited their utility for general spatial data handling. Current approaches tend to make use of a general DBMS for handling the spatial attribute information and specialized software for storage, retrieval and manipulation of the spatial data. Arc Info is a good case in point since it consists of Info (a commercial, semi-relational DBMS) and Arc (a specialized spatial data handling system).

The inability to enlist DBMS to efficiently handle large volumes of spatial data presents a real obstacle in the development of global databases. Similar impacts are found in the image processing field, where picture data management is also of serious concern.

4. Software engineering

Within the last decade increasing attention has been given within computer science to the problems of efficient design of large software systems. This work has become known as software engineering and, through the concept of the system life cycle, has led to the development of conceptual models and tools for effective system design. This work was badly needed because of the large number of system disasters that occurred in the late 1960s and 1970s. Many systems failed; and the reason for most of these failures was determined to be bad design: systems were over-budget and over-schedule and failed to work as desired.

The same problems had, of course, plagued the area of geographic information systems. Many of the early systems were held to be failures due to poor performance and, in some cases, the offending systems vanished from the scene. Other early systems managed to survive, often through a combination of good luck as well as sometimes good design. Attention was given in the GIS area to problems of system design and selection at an early date, and it is interesting to note that many of the notions contained in these early design models parallel concepts found in modern software engineering practice (e.g. structured functional requirements analysis).

Structured design approaches are becoming more common in the spatial data handling area and initial attempts are being made to construct the types of engineering cost estimation functions that are found today for less specialized, large software systems. The tools of software engineering are also being applied to the development of more efficient structures within individual segments of GIS operation (manual digitizing, for one).

5. Remote sensing and photogrammetry

In a sense, the great majority of the data contained in digital, spatial databases is derived from remote sensing. The derivation is indirect since most data are captured by digitization (either manual or automatic) from map documents, which are, in turn, frequently derived from photogrammetric processing of aerial photography. However, the direct utilization of remote sensing inputs (especially those based on orbital sensors) is found in only a limited number of cases at present.

The reasons for this limited interaction appear to lie in misperceptions by both groups (remote sensors and GIS managers) as to the nature of the data created by remote sensing systems and used by geographic information systems. GIS managers, who are used to dealing with map data, which normally carries no information pertaining to the accuracy or precision of individual elements (and hence is interpreted as being highly precise), view remote sensing data as relatively inaccurate and hence of limited utility. Those organizations generating remote sensor data have neither understood this view, nor have they devoted much attention to the comparative economics of the two data sources (maps and remote sensors).

This lack of interaction between GIS and remote sensing systems is indeed unfortunate since significantly higher levels of interaction would improve the effectiveness of a GIS through the availability of more current data, and would improve the quality of remote sensor data through utilization of ancillary data contained in the spatial databases of the existing geographic information systems.

Information derived from satellite stereo-pairs through digital photogrammetry has become one of the important inputs to GIS for further processing, like contour mapping, DTM and other inputs.

D. Development areas in GIS

Spatial relations theory: There is, at present, no coherent mathematical theory of spatial relations. This lack seriously impedes both the quality of existing research and the speed at which developments can take place. The impact of this lack is especially felt in the attempts to develop true spatial database management systems and in the creation of efficient algorithms for spatial data handling. Similar impacts are felt in the areas of image processing and computer vision.

Artificial intelligence: Important developments are occurring in the field of artificial intelligence and will certainly have spatial data handling applications. The developments should be watched closely by the spatial data handling field and applied to spatial data handling technology as promptly as possible.

Expert systems: Expert systems that include spatial data handling capabilities are likely to attract new users to the spatial data handling field. The development of expert systems helpful to users handling spatial data should be encouraged.

Database queries: It is extremely difficult to query large spatial databases, but it is extremely important to be able to do so and do so efficiently. Research and development in this area, in addition to the basic theoretical work on spatial relations, is especially important for the creation and use of very large or global databases.

Global databases: A series of concerns must be addressed if successful global databases are to be created for wide public use. A variety of problems dealing with creating a global database will need to be addressed simultaneously if progress is to be made. Pilot studies of global databases need to be undertaken.

Improved data input: Data input is probably the biggest bottleneck in spatial data handling systems at present and represents the greatest single cost in most projects, especially when the database is very large. Research and development in a number of related areas may need to be done before progress in automation can be significant; these areas probably include feature recognition, cognitive science, artificial intelligence and others. Documentation of present methods, costs and throughputs needs to be obtained to provide baseline data against which potential improvements can be measured.

Data updating: Improved methods for updating data in spatial databases are needed. Continuing development of updating methods is of great importance to the integrity of databases and to the maintenance of user confidence in spatial data handling. It is here that remote sensing inputs can be of substantial value.

Benchmarking: Benchmark tests are useful in measuring the performance of a wide variety of spatial data handling functions, and should probably be more widely used in selecting systems. Persons expert in the design and conduct of benchmark tests should be urged to share this knowledge. Some publication of benchmark results would be useful to potential system users, if accompanied with appropriate cautionary remarks.

User-friendliness: Spatial data handling systems need to be user-friendly. Serious (as opposed to cosmetic) attempts at creating user-friendly geographic information systems need to be continued, using the best guidance available from a wide range of fields, such as ergonomics, cognitive science and others.

Need for improved efficiency: There is a need for improved efficiency in nearly every aspect of the spatial data handling function. Efforts in software development, algorithmic analysis, database structure design, ergonomics, engineering economics and a whole range of other areas are needed in order to achieve improved efficiency in spatial data handling functions.

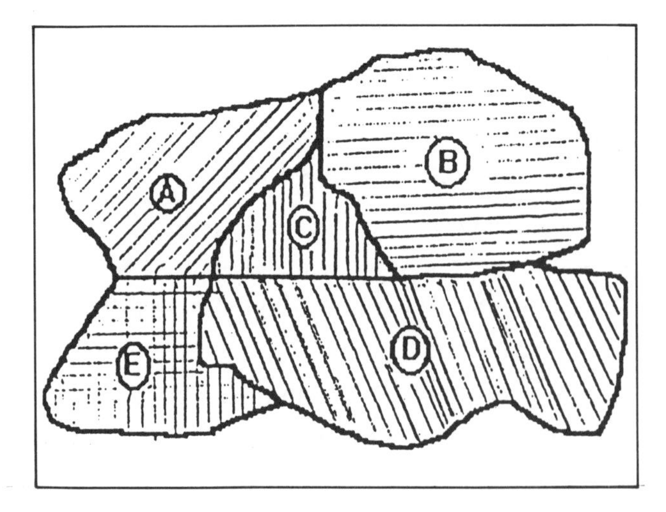

Figure 1. An example of a choropleth map

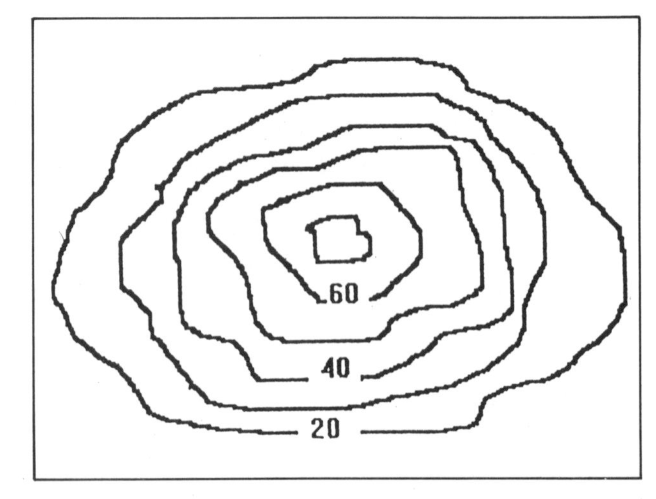

Figure 2. An example of an iso-line map

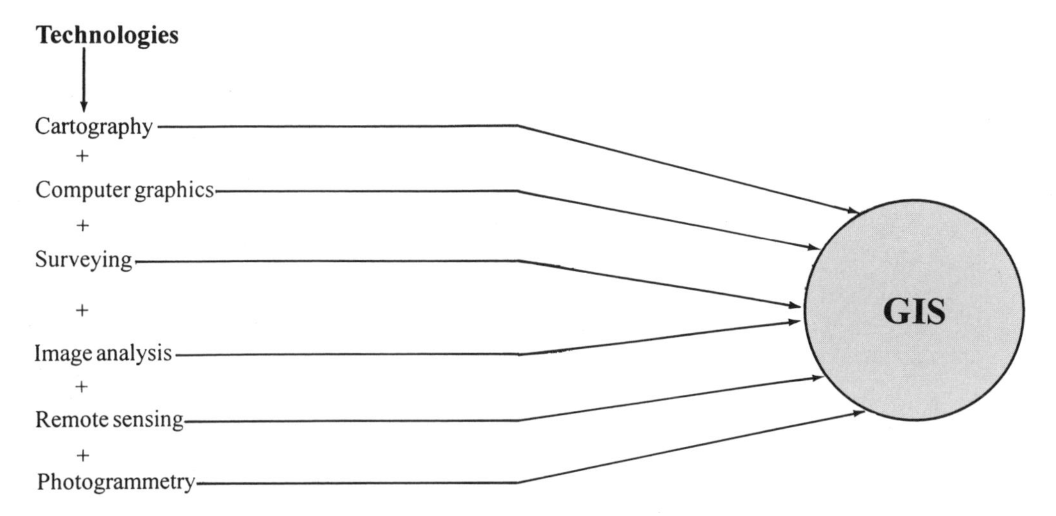

Figure 3. Other technologies embedded in GIS

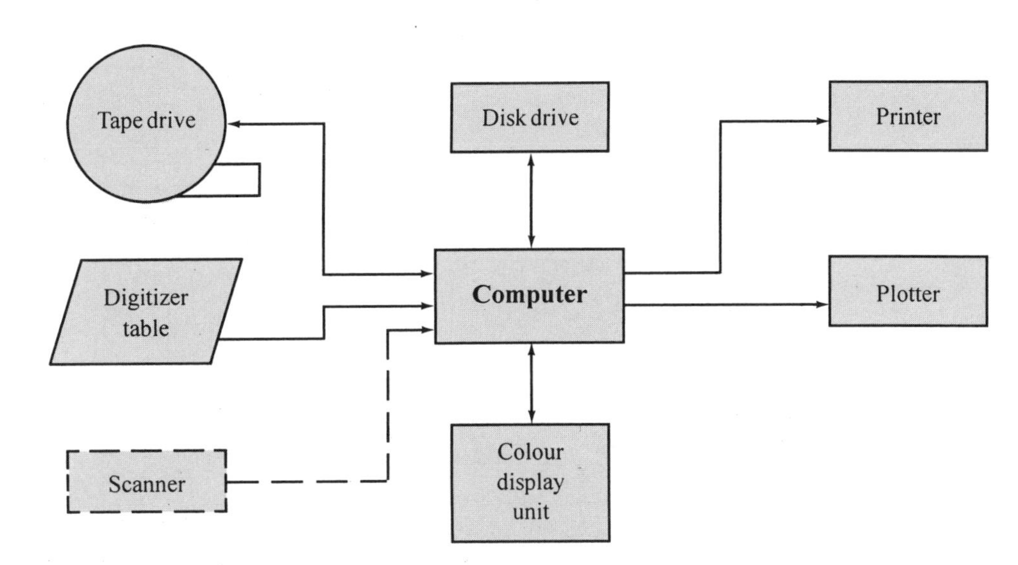

Figure 4. The major hardware components of GIS

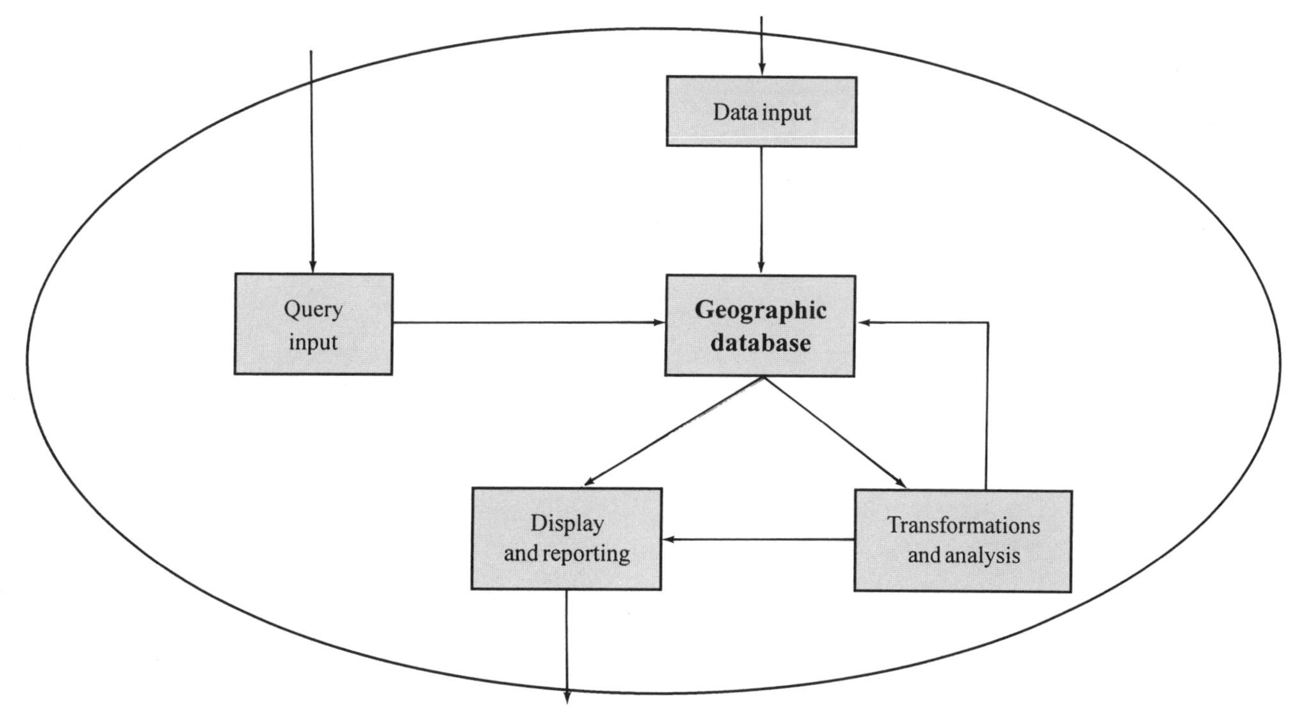

Figure 5. The main software components of GIS

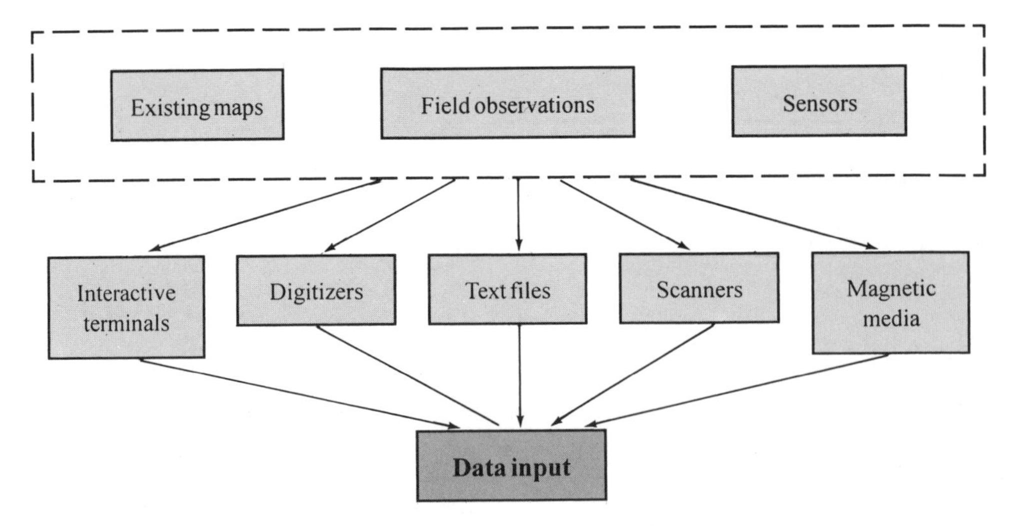

Figure 6. Sources of data input

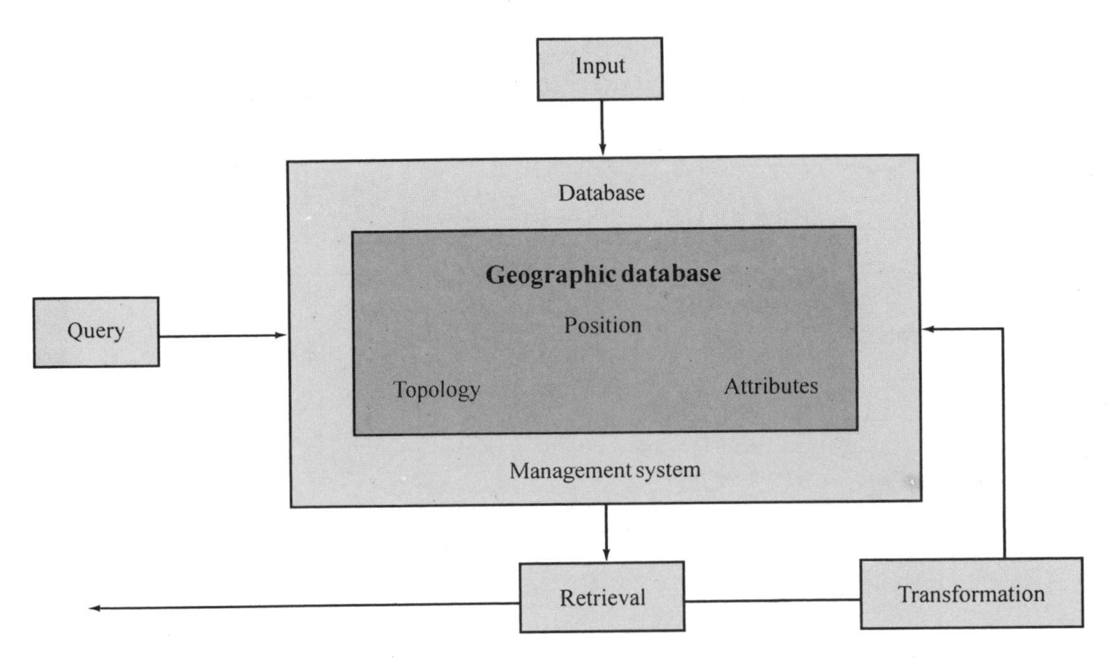

Figure 7. Data storage and database management system

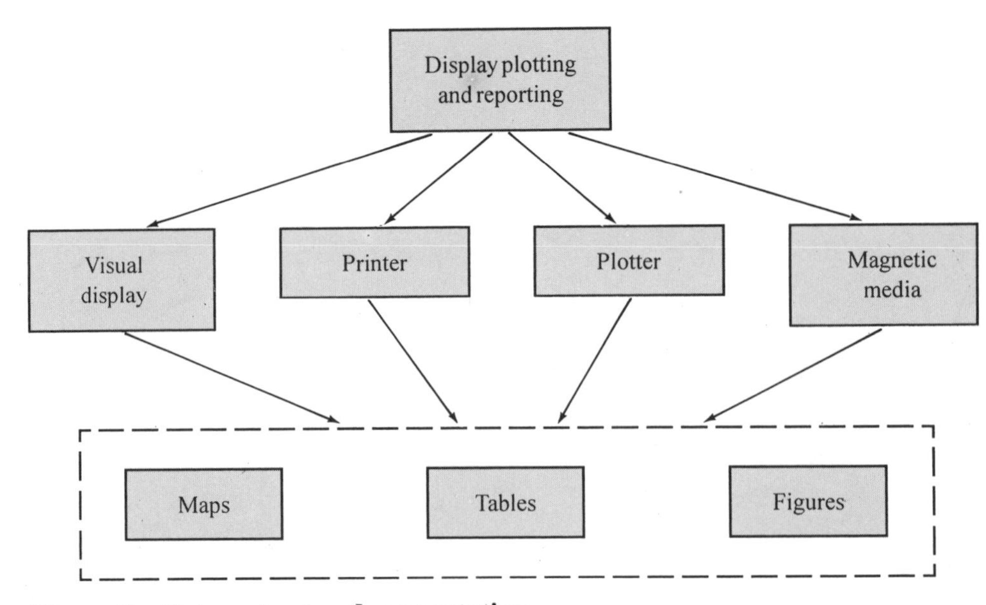

Figure 8. Data output and presentation

SPATIAL INFORMATION INFRASTRUCTURE FOR INDIA: THE NRIS PROGRAMME

*Mukund Rao**

A. Introduction

India has been one of the major users of GIS, specifically in combination with the use of satellite data from its indigenous Indian Remote Sensing satellite (IRS).

IRS data is used in national-level projects in different resource management areas, and the remote sensing technology has matured to cover diverse resource themes/areas such as forestry, wasteland mapping, agricultural crop acreage and yield estimation, drought monitoring and assessment, flood monitoring and damage assessment, land use/land cover mapping, wasteland mapping, water resource management, groundwater targeting, marine resource surveys, urban planning, mineral targeting and environmental impact assessment and other themes.

Recently, India has taken up a massive programme for the generation and implementation of locale-specific integrated land and water resource development plans through the Integrated Mission for Sustainable Development (IMSD). The IMSD approach involves integration of thematic information on various natural resources -- land use/cover, types of wastelands, forest cover/types, surface water resources, drainage pattern, potential groundwater zones, geomorphology (land-forms), geology (rock types, structural details, mineral occurrence) and soil types -- derived from satellite data and integrating the information with other ancillary information, meteorological and socio-economic data in a geographic information system (GIS) environment to arrive at locale-specific prescriptions for development. Based on the interaction among the basic resources of land, water and vegetation, which form the major components of the primary production system and "carrying capacity" of the region, useful inferences are drawn about their predicted behaviour in meeting the various planning goals.

These developmental activities require a systematic and comprehensive spatial information system, around a GIS core, and containing both thematic maps and non-spatial data so as to be amenable to integration.

Towards this, a major effort to organize a spatial information infrastructure for the country has been taken up as part of the National Resources Information System (NRIS).

The objectives of the NRIS programme are:

(a) To systematize, design and establish sectoral and hierarchical information systems for the totality of natural resource management in the country;

(b) To design information transfer mechanisms for a systematic networking of the sectoral and hierarchical information systems;

(c) To develop appropriate technology and establish an infrastructure base for hardware, software and trained manpower for the operational usage of NRIS;

(d) To evolve an organization structure for the management and upkeep of the NRIS nodes in an operational, scenario.

The NRIS is a three-tiered system catering to policy making, planning and implementation, as shown in figure 1, and has two major elements:

(a) Information Generation System (NRIS-IGS), for acquiring and generating data on natural resources. This will include spatial data, from both remote sensing and conventional surveys, and

* Indian Space Research Organization HQ, Bangalore, India.

non-spatial data from census and other sources. The remote sensing segment forms a major part of the NRIS-IGS as much of the natural resources maps/information -- land use, forests, surface water, structures, geomorphology and crops -- can be generated using remote sensing data. Therefore, remote sensing is a major and important element for NRIS-IGS;

(b) Spatial Information System (NRIS-SIS), which pertains to the systematic organization of a natural resource database; multi-parameter integration of spatial and non-spatial data sets; and information presentation in an optimum manner, oriented towards generating scenarios of sustainable development.

The NRIS would be a set of natural resource databases organized in two hierarchies, as follows:

(a) Centre-state-district, catering to resources and sectors requiring integrated resource management or for preparing plans according to administrative units -- district plans, state plans, or central plan. For this hierarchy, there would be approximately:

- 520 + NRIS-district nodes;
- 26 NRIS-state nodes.

(b) Centre-region-project, catering to sectoral resource management needs (for example, water resources or forestry) where the natural resource boundary (forest unit, watershed and so on) is the unit. For this hierarchy there would be:

- 200 NRIS-project nodes;
- 48 NRIS-regional nodes.

Figure 2 shows the framework of NRIS in the context of the two hierarchies. The NRIS nodes would have to be interlinked/networked so as to cater to the free flow of information to the next higher level in the hierarchy. This networking would be both horizontally (intra-node) and vertically (inter-node).

B. Some NRIS applications

A brief overview of some of the projects that have been completed is given below:

(a) NRIS-Bharatpur: A pilot-project for establishing a regional planning system for the district of Bharatpur, Rajasthan, has been executed by the Department of Space (DOS) and the Town and Country Planning Organization (TCPO), New Delhi, and with the active participation of the District Collectorate, Bharatpur. The GIS database has been based on the SOI map base at 1:50,000 scale and consists of twenty-nine spatial elements -- mainly different types of map at 1:50,000 scale obtained from remote sensing data and other conventional sources. The non-spatial data includes village-wise census data of 1981 and 1991. The core of the analysis has been:

(i) Intra-district disparities towards identifying the imbalances in development where demographic indicators have been studied, based on the pattern of population, SC/ST composition and their occupations: occupational structure indicators, such as participation ratio of the population in various socio-economic activities, and overall village development indicators and infrastructural facility indicators, such as education, medical/health, general facilities, communication and road/transportation network;

(ii) Land capability assessment where two major aspects of the land have been assessed -- multi-date remote sensing data-based land use (and the changes therein for the whole district) and land capability -- based on an integration of the soil information, slope categories and the land-use classes;

(iii) Analysis for agricultural development, based on an indexing of the village-wise data on irrigated area, irrigation facilities, agriculture occupation and agriculture facilities. The Agriculture Development Index is then integrated with land capability to generate village-wise developmental needs for the agriculture sector;

(iv) Functionality assessment of settlements based on service functions -- education, medical, transportation/communication and general facilities -- available in each village and identification of service development needs of villages;

(v) Village accessibility analysis, based on the nearness of each settlement to a road and its accessibility distance. The villages have been categorized based on population and access to identify developmental needs for roads.

(b) NRIS-National Capital Region (NCR): With a view to develop core modules of a GIS and also address application aspects -- analysis for the growth profile of urban areas in a sub-region of NCR, urban land suitability and land-use changes over a five-year time-frame -- a photo-type project has been done for NCR.

Presently, the NCR Planning Board (NCRPB) is in the process of revising the NCR-Regional Plan for 2001. Towards this, a spatial information system for the national capital region around a GIS core is being organized. One of the objectives of the project is the establishment of the NCR-GIS for operational use. The NCR-GIS is to be established for the total region, which covers about 40,000 sq km. The NCR-GIS caters to macro-level applications at 1:250,000 scale with specific applications for urban growth profile analysis, land-use change analysis, land suitability analysis, and environmental sensitivity analysis. Later extension to meso- and micro-levels at 1:50,000 and 1:10,000 scales for smaller regions in the NCR is also planned. ISRO has already designed the NCR-GIS database, and the organization of the database is almost complete;

(c) NRIS-wasteland development: Towards identifying reclamation alternatives for wasteland, DOS has completed GIS-based integrated analysis in three districts -- Dungarpur, Rajasthan; Purulia, West Bengal; and Sundergarh in Orissa. For example, in Dungarpur, spatial patterns of the village-wise needs assessment have been generated to serve as indicators and provide inputs towards the planning process. The resource assessment is based on the integrated analysis of the physical characteristics of the land and identification of land units having homogeneous or similar characteristics. For each homogeneous land unit, a set of preferred uses have been identified. Based upon these assessments, the implementation plan for cluster of villages has been generated;

(d) NRIS-district-level planning: A project on district level planning was taken up for Panchmahals District, Gujarat, with the aim of aiding the district planning process using GIS techniques. The database was organized at 1:250,000 scale and included land-use maps, soil maps, groundwater maps and wasteland maps. Non-spatial data included the village-wise census data. The database has been organized at 1:50,000 scale. The following analysis was carried out:

(i) Slopes, groundwater prospect, land use and soils were integrated to derive a composite land development unit (CLDU) map. For each CLDU, a prescription for soil conservation measures has been made based upon its characteristics;

(ii) Service centre hierarchy has been determined for the district by identifying the cluster of consumer settlements around each centre and based on the criterion of minimum radial distance;

(iii) Spatial distribution of demographic status and transport/communication infrastructure has also been assessed.

(e) NRIS-mineral targeting: The Vasundhara project, a collaborative project between the Geological Survey of India (GSI) and DOS, is an integrated appraisal of data from satellite remote sensing, geophysics (airborne and ground) and geological ground truth for mineral targeting on a regional scale. The area of investigation is about 400,000 sq km, forming a major part of the south Indian peninsular shield. Integrated appraisal of geological interpretation of Landsat TM imagery, airborne geophysical data (magnetic, electromagnetic and radiometric) and inventory of all mineral occurrences have resulted in preparation of various thematic maps and have led to delineation of numerous structural features, demarcation of numerous granite bodies, tracing of the extension of greenstone belts, and identification of numerous circular features. As a spin-off product, derivative maps have been prepared on intrusive complexes, dykes and sills, and classification of granites and ornamental stones. Ten areas with potential for various types of mineralization such as base metals, diamonds, rubies, gold and chromite have been delineated.

C. Conclusions

The organization of a systematic infrastructure of spatial databases for the country is an essential need to support decision-making and developmental activities. While the canvas of NRIS is a vast one, initially the NRIS programme has taken up the development of 30 district and four state nodes in the country, with plans for extension to other districts/states in a phased manner.

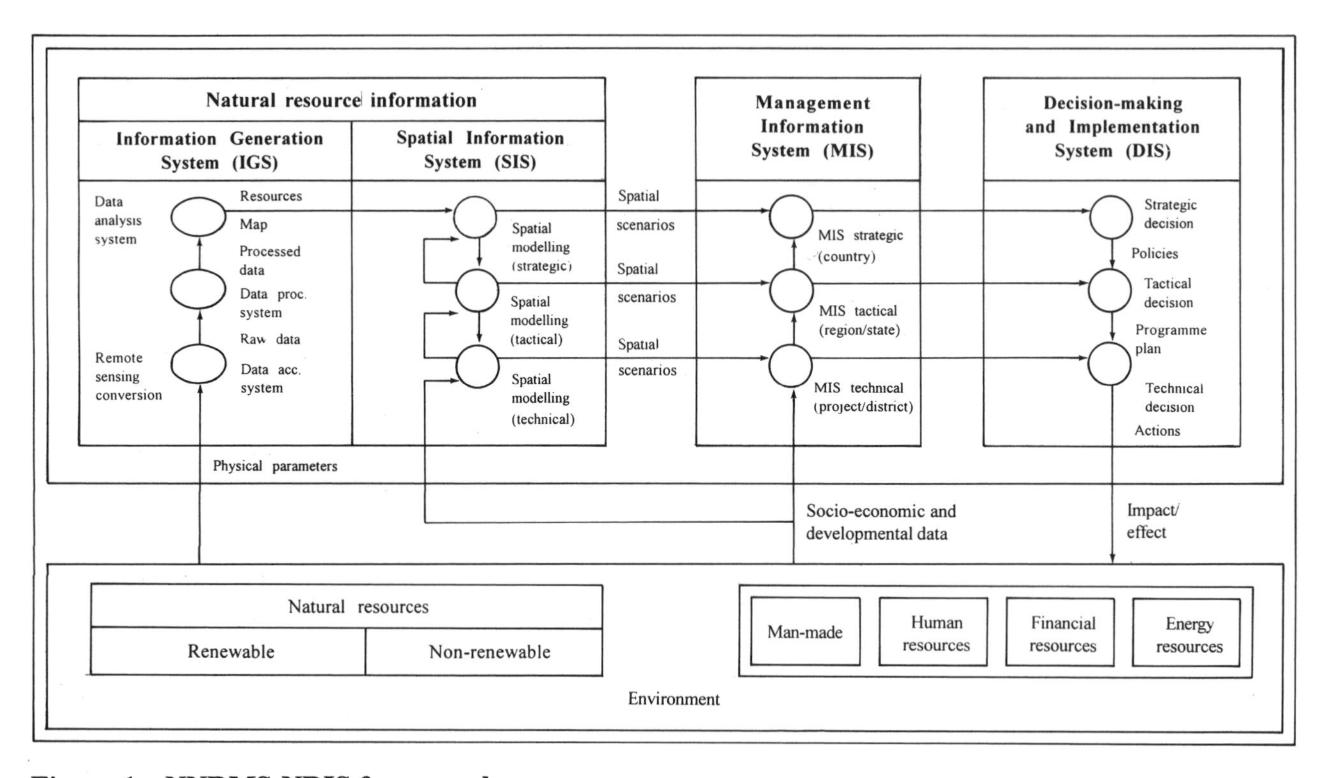

Figure 1. NNRMS-NRIS framework

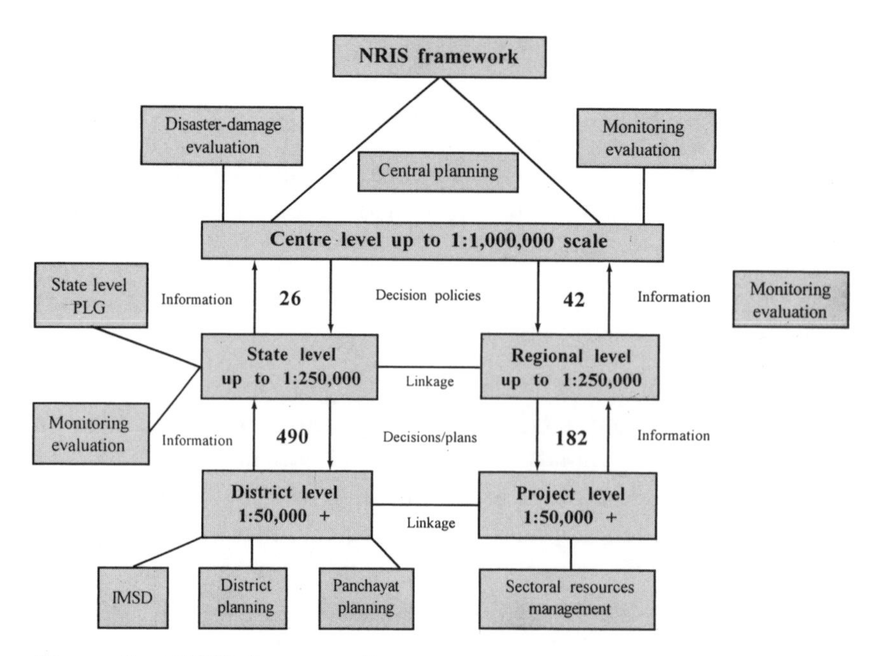

Figure 2. NRIS framework

FUNDAMENTALS OF GIS AND STANDARDIZATION ISSUES

*Mukund Rao**

A. Introduction

A geographic information system (GIS) has two distinct utilization capabilities -- the first pertaining to querying and obtaining information and the second pertaining to integrated analytical modelling. However, both of these capabilities depend upon the core of the GIS database that has been organized. Many a GIS utilization has been limited because of improper database organization. The importance of the GIS database stems from the fact that the data elements of the database are closely interrelated and therefore need to be structured for easy integration and retrieval. The GIS database has also to cater to the different needs of applications. Some of the fundamental aspects of GIS and issues for standardization are discussed in this paper.

B. Data in GIS

Broadly categorized, the basic data for any GIS application have two components:

(a) Spatial data -- consisting of maps that have been prepared either by field surveys or by the interpretation of remotely sensed data. Some examples of the maps are the soil survey map, geological map, land-use map from remote sensing data, village map and so on. Many of these maps are available in analog form, and recently some map information has become available directly in digital format. Therefore, the incorporation of these maps into a GIS depends upon whether it is in analog or digital format -- each of which has to be handled differently;

(b) Non-spatial data -- attributes that are complementary to the spatial data and describe what is at a point, along a line or in a polygon, as well as socio-economic characteristics from census and other sources. The attributes of a soil category could be the depth of soil, texture, erosion or drainage, and for a geological category they could be the rock type, its age or major composition. The socio-economic characteristics could be the demographic data, occupation data for a village or traffic volume data for roads in a city, among others. The non-spatial data is mainly available in tabular records in analog form and need to be converted into digital format for incorporation in a GIS. However, the 1991 census data is now available in digital mode, so direct incorporation into GIS databases is possible.

The issues asociated with data are:

(a) *Formats* of these spatial and non-spatial data sets are different and vary considerably. Even paper maps are of different sizes and scales, so handling these by manual means is difficult. Because of the variation in formats, the organization of a GIS database has to be initiated with the step of standardizing the format and bringing the data sets to the same format and scale. However, the problem would be acute with analog non-spatial data, as the format of tables, their contents and coding schemes would be varying. With the volume of non-spatial data (say, of villages, districts, states and country) being large, the effort at standardizing the formats is Herculean;

(b) Furthermore, the issue of inputting data is compounded by the *differences in classification* schemes and legending associated with features in the paper map. With the variation in classification schemes of data sets, the understanding and perception of users also varies and this brings in a level of confusion in the application;

(c) *Source of spatial and non-spatial data* is an important issue as it brings about the details of the data collection activity and helps identify the need for data generation. Spatial data sets are mainly obtained through remote sensing or conventional surveys.

* Indian Space Research Organization HQ, Bangalore, India.

Remote sensing data is a classic source of data on natural resources for a region and provides a record of the continuum of resource status because of its repetitive coverage. Remotely sensed data in the form of satellite images can be used to study and monitor land features, natural resources and dynamic aspects of human activities and towards preparation of thematic maps depicting the status of various resources. Therefore, remote sensing data has to be an important element for the sustainable development activity, and how this can be done is explained in the ensuing section. Remote sensing data provides information on (i) land -- land use, forests, urban growth, crops, (ii) water resources -- surface water bodies, groundwater prospects, recharge site, (iii) oceans -- biological parameters of oceans like chlorophyll, phytoplankton and fish schools, and physical parameters of oceans, and (iv) atmosphere -- clouds, aerosols and minor constituents, and atmospheric chemistry. The synthesis of these information sets can provide global change scenarios and also locale-specific prescriptions for sustainable development.

Most of the spatial data or thematic maps are available from the central and state survey agencies. Much of the non-spatial data comes from the census department, municipalities, survey agencies and other bodies;

(d) *Age of data* defines the temporal domain of the database, making it either useful or useless for a particular end application. For example, if the application is to study the impact of pollution in an urban area then the pollution data needs to be current, and the use of past data would render the impact analysis ineffective.

C. GIS package

The GIS package is the core of the database as both spatial and non-spatial databases have to be handled. The GIS package offers efficient utilities for handling both these data sets and allows for (a) the spatial database organization, (b) non-spatial database organization, mainly as attributes of the spatial elements, (c) analysis and transformation for obtaining the required information, (d) obtaining information in specific format (cartographic quality outputs and reports), and (e) organization of a user-friendly query system. Different types of GIS packages are available and the GIS database organization depends on the GIS package that is to be utilized. Apart from the basic functionality of a GIS package, some of the crucial aspects that impact the GIS database organization are discussed below.

D. Data structure of the GIS package

The data structure of the GIS has a major bearing on the accuracy of the representation of spatial data. Most GIS packages adopt either a raster or vector structure, or their variants, internally to organize spatial data and represent real-world features. The key element in both these models is that links are established between attribute information and spatial features. The precise techniques used to create these links vary from GIS to GIS but, in general, involve establishing a pointer between each spatial feature in the database and its associated attribute information. In the raster model, the links are implicit in the way that specific attributes are assigned to individual map layers, by a process of assigning specific attribute values to the spatial entities (i.e. cells) in each layer. In the vector model, the links are established by arranging for each spatial feature's unique identifier or ID to be recorded in a key field of the appropriate attribute table in a DBMS that stores its attribute information. The way spatial data is organized and handled in the two structures and the advantages/disadvantages of each is discussed below:

(a) *Raster GIS packages* organize spatial features in a spatial domain of grids/rasters, which are defined by dividing the space into equal units and obtaining a matrix of grids/rasters. Rasters are limited by the area they can represent and also the limits of storage space. Also, the fineness of data is limited by the cell size -- thus the area of coverage is traded off with the resolution of the coverage. The storage problems are handled by resorting to coding such as run-length coding, chain coding, block coding and other methods. The capabilities of a raster-based cartographic modelling system ultimately arise from the functions associated with individual data-transforming operations and the way in which these operations are combined. This transformation of data is facilitated by the fact that map layer zones are represented not by lines or symbols but by numerical values. It is also facilitated by the fact that these values are directly associated with individual locations. The use of numbers here makes it possible to transform geographical characteristics using mathematical and arithmetical functions.

The basic problem here is the size of the grid/rasters, which limits the level of accuracy and, once organized at a size, does not permit the representation of features at finer resolutions. For example, a 10-m grid would mean achieving an areal accuracy of 100 sq m, so representing features smaller than 10 m would not be possible;

(b) *Vector GIS packages* represent spatial features in a very accurate manner and adopt sophisticated topological relationships to represent different features. The vector data structure represents each geographical feature by a set of coordinates. Vectors as xy coordinates define points, lines and polygons. The basic premise of the vector-based structuring is to define a two-dimensional space where features are represented by coordinates on the two axes. Generally, representing points and lines is straightforward -- points are characterized by an xy coordinate pair, and lines by a set of xy coordinate pairs with a specific beginning and ending vector. However, representing polygons in vector storage poses a challenge. Modern vector-based structures overcome this non-optimal storage by building and defining a topology -- characterizing relationships and adjacency of features. The premise of topology is based on a conceptual definition of relationships between points and lines and lines and polygons. A vector -- one xy coordinate pair, indicating a specific location -- is the basic unit and all geographic entities, i.e. points, lines and polygons, can be defined from the basic definition of vectors. A point is related on a one-to-one basis with vectors, so points are represented by one xy pair. A line is characterized by a series of vectors and are characterized by one-to-many relationship with vectors. A polygon is represented by a set of lines -- thus forming a closed area -- and so has a one-to-many relationship with lines. In this manner, a logical hierarchy of vectors-points-lines-polygons can be constructed systematically, thus reducing redundancy as shared lines are stored only once. Vector structures would require more storage space and also more processing to be able to determine the topology and store the points, lines and polygons in an effective manner. The overall processing algorithms for vector structures are also complex.

E. Raster versus vector: the debate

As there are two characteristically different ways of spatial data representation for GIS applications, questions of choosing one method over the other arise. Some of the issues that need consideration are as follows:

(a) Vector data is precise and has no approximation errors for measured quantities like area, length or perimeter (subject, of course, to the representational accuracy of the hardware). Owing to discretization, raster data suffers from such approximation errors;

(b) Generally, raster data has higher storage requirements, although compression techniques can be adopted. Vector storage is relatively smaller;

(c) GIS manipulation operations on raster data are computationally faster and less complex than their equivalents on vector data. Imposing a structure on 2-D vector data that is sufficiently formal and supports efficient searching is fairly complex;

(d) Raster data is not easily amenable to association of attribute data with spatial features such as points, lines or polygons. This is due to the fact that the basic entity in the raster approach is the grid cell, and entities such as points, lines and polygons are not recognized as objects in their own right;

(e) Raster is limited by the cell size, and sampling or disaggregation to a lower size is not possible. Vector structure does not impose this limitation so analysis, plotting and other operations are more accurate in vector structure.

As spatial data is available in both raster and vector forms, a GIS package cannot ignore one form of data in preference to the other. Generally, GIS packages choose one of the two forms of data representation as the primary method of representing spatial data and provide conversion utilities from one form to the other.

F. Attribute data management

Attribute data is an essential element of a GIS database, and the GIS package must have the capability to handle attributes efficiently and also in conjunction with spatial data. Attribute data are best handled in a database management system (DBMS), which manages the attribute data as tables with tuples representing one entry and fields representing the attribute characteristics. Most GIS packages have linkages to DBMS packages like dBase, FoxBase, SQL databases and so on. The DBMS packages are embedded into the GIS package and through pointers and keys relate the attribute data and the spatial data. However, some GIS package utilize file management systems to organize the non-spatial data. The utility of a GIS is also determined by the DBMS interface it has.

G. Functionality of the GIS package

The functionality of a GIS package is characterized by the range of operations possible using the GIS. The functional character of GIS packages varies considerably but is mainly based on a set of core modules that perform similar spatial and non-spatial data-handling operations. The functionality of GIS packages can be classified into three categories:

(a) Data entry, editing and maintenance, which allow for the entry of spatial and non-spatial data into the GIS database and also organize and maintain the database internally. Furthermore, these modules allow editing operations in terms of adding, moving, deleting and updating of features and also organize the topology, coding of rasters and so on;

(b) Analysis/transformation/manipulation, which allow for the multi-parameter integration of spatial and non-spatial data sets in the framework of an application model;

(c) Data display and output, which allow for displaying and outputting the data either as a spatial output or as a non-spatial table or report. They also allow for the querying of data on the screen.

Some of the critical parameters to look for in a GIS in terms of functionality are given in table 1.

H. File formats of GIS

File and data organization in GIS is an activity that is taken care of by the GIS package itself and is an internal physical organization of files. Each GIS package has its own file system organization, which could be either a single file or a set of files, and it is transparent to the user. Generally, the internal organization of files and their formats is proprietary to GIS packages and is not available to users. However, to take care of data exchange, gateway utilities are provided by each package.

I. Gateways to GIS packages: essential for data exchange

Because of the variation in the file formats of GIS packages, gateway utilities are critical for data exchange from one GIS to another GIS package. These are basically "translator" routines that convert the proprietary formats to a readable, free format and can also convert the free format to proprietary. The gateway modules are an essential feature of any GIS.

Table 1. Broad requirements of GIS packages

Module	Description/functionality
Spatial input	Ability to handle points, lines and polygons with user-specified names/labels; ability to handle z-axis of data; also user-definable coordinate system; polyconic projection. Accurate representation of coordinates, length and area estimates. Interface to digitizers, scanners, frame-grabbers, remote sensing systems. Gateway to remote sensing analysis systems, other GIS packages.
Tabular input	Embedded DBMS package for handling all non-spatial tables; all DBMS functions for table creation, editing and updating, joining, data listing, report generation, views and so on. Embedded attribute table generation for spatial features. Keyboard and file interface for input. Gateway to other DBMSs -- dBase and FoxBase.
Spatial edit	Ability to pick spatial features and perform move, copy, delete, rotate, split, extend and add features. Automatic polygon generation and updating of point, line and polygon attributes in tables. Ability to report editing errors of over-shoots, under-shoots, polygon topology errors, and removal of errors -- snap or coalesce features, remove over-shoots, label errors and so on.
Integration	Spatial overlay -- union, intersect and other geometrical variations; attribute merging also must happen. Modules to remove slivers, generate buffers; aggregation and abstraction, selective copying, distance and nearness estimation, geometrical transformation, generation of slopes, aspect and other 3-D planar derivatives. Non-spatial data analysis -- mainly statistical estimates of sum, average, frequency, standard deviation. Generate flat-files of data for analysis in other packages.
Symbolization	Point, line and polygon symbolization using colours and marker/line/hatch patterns; selective symbolization. Free text placement and annotation with fonts, colours and size specification; text from attribute table as labels. Library of standard symbols; user-defined symbol library generation and use.
Output	Generation of near-cartographic quality output by combining various spatial and non-spatial features on screen as a composition process. Legending, annotation, map symbol placement -- boxes, grids, scale bars, direction arrows. Edit outputs to add, delete, remove elements in composed maps. Interface to plotters, printers with scaling facility.
Query	Spatial queries -- point-and-ask on screen and obtain attribute information. Logical queries -- ask-and-show on screen and obtain spatial features. Query from attribute tables and logical views of other tables. On-screen queries of coordinates, length, areas, attribute information and other data.
Interface	Easy-to-use and learn graphic user interface; mainly menu-driven; vernacular language interface option.
Others	Database maintenance modules -- backup, restore, copy, delete, rename and so on. Security against system crashes; password-protected databases; operation logs.

J. User interface: standards of GUI

User interface in a GIS is a very critical element as it provides the front end for the user to interact with the package. Users of GIS packages would find it inhibitive if the interface were complex and demanded a lot of technical competence and understanding for using the package. Command interface, though elegant, suffers from the fact that a user has to remember the command and its syntax, which puts a strain on the user. A state-of-the-art GIS package would have almost 200-250 commands, and remembering these would be undesirable and difficult.

The power of a graphical user interface (GUI) has been widely appreciated and recognized, where the interface depends more on graphic metaphors, rather than syntaxes. The GUI makes use of the Windows environment on PCs and workstations to provide pleasing and easy-to-understand interfaces. The GUI can be thought of in the same way as a "dashboard". Just as a car contains a window, steering wheel, gas pedal, brakes and so on, an on-screen GUI dashboard contains objects such as a viewing screen as well as various types of status lines, menu bars and icons. This on-screen GIS dashboard provides the user interface for the GIS, so what-you-see-is-what-you-get (WYSIWYG) becomes easy to use and it is easy to understand the operation.

Standardizing the GUI is an essential element so that a common graphic metaphor is developed as a core. This would make the user's understanding much easier.

K. Selecting a GIS: benchmarking

The process of evaluating a geographic information system package is one of the prime activities for any organization or agency interested in GIS procurement. The evaluation process is a difficult one because of the complexity of the GIS technology and because of a variety of techniques and methods adopted for realizing the various functions of the packages. Moreover, comparison of the techniques and methods used for the functions is not possible because of the non-availability of this information from the vendors. Moreover, there is a large number of GIS packages available in the market -- making it almost impossible to perform a systematic evaluation. There are benchmarks for evaluating GIS packages, which have been developed by various groups. ISRO has developed and adopts a GIS benchmark called GISMARK and this has been used to evaluate the performance of different GIS packages. GISMARK is a 26-step procedure consisting of mainly spatial integration functions and is developed around the core of an application model involving the spatial integration of different elements, with the major aim being to find suitable land for urbanization. The application model is based on a well-defined procedure for data integration and analysis which has already been applied for a major project.

The evaluation from a benchmark, like GISMARK, is an important aspect. The evaluation is oriented towards evaluating:

(a) Performance, in terms of time taken to execute the benchmark -- in individual steps and in total -- which would have a major bearing on the GIS selection;

(b) Functional compliance, in terms of the successful completion of each step of the benchmark. This compliance evaluation will indicate the capability of the GIS to perform integrated analysis similar to that of the benchmark model;

(c) Accuracy of theme attributes, in terms of the accuracy of results obtained from the benchmark operation -- area and perimeter of polygons, length of lines, number of features created and so on. For example, the accuracy of results obtained from a raster GIS will be limited by the raster size, and this could be limited by the software or system capabilities. However, a vector GIS may bring out different results because of algorithmic differences;

(d) Accuracy of results, in terms of the final result of suitability for urbanization. This accuracy is evaluated on the basis of evaluating the suitability characteristics, areas under different categories and other elements. This is an important measure of the application model and thus will be an assessment of the ability of the GIS to implement such models;

(e) Result correspondence, in terms of the variation in the results obtained from the GIS packages under evaluation and the standard result expected from the benchmark. The standard result from the benchmark is generally provided by the benchmark.

The GISMARK benchmark provides scope to evaluate the GIS performance on the above parameters, plus additional parameters in terms of evaluating space, ease of operations and so on.

L. Standardization issues

Generally, a proper database organization needs to ensure the following:

(a) Flexibility in the design to adapt to the needs of different users;

(b) A controlled and standardized approach to data input and updating;

(c) A system of validation checks to maintain the integrity and consistency of the data elements;

(d) A level of security for minimizing damage to the data;

(e) A way to minimize redundancy in data storage.

While the above are general considerations for database organization, in a GIS domain the considerations are more pertinent because of the varied types and nature of data that need to be organized and stored. As in any normal database activity, the GIS database also needs to be properly designed to cater to the needs of the application that proposes to utilize it. The design would also define a comprehensive framework of the database and allow the database to be viewed in its entirety so that interaction and linkages between elements could be pre-defined and evaluated. The design would also permit identification of potential bottlenecks and problem areas so that design alternatives could be considered. The design would also help in identification of the essential and correct data elements to be incorporated into the database and filter out irrelevant data. Definition of updating procedures could also be a part of the design activity so that newer data can be incorporated in future and the database is "active".

M. National or organizational commitment

An important aspect of any nation or organization setting up GIS-based information systems is the level of commitment of the nation or organization. This involves a decision at the national or organizational level committing itself to the GIS and also defining the GIS as an amalgamated system of the organization and its work culture. This would require an orientation/training of senior decision makers with specific emphasis on GIS and its capabilities, differences between GIS and other packages being used in the organization, what can be achieved by the organization using GIS, implications of GIS to the organization's goals and objectives, assessment of effort, broad system definition and cost/time/manpower implications.

N. End use and level of GIS

It is important to specify the ultimate use of the GIS database in a single statement. Some examples could be "GIS database for urban planning at micro-level"; "GIS database for water supply management"; "GIS database for wildlife habitat management"; "GIS for sustainable district development" and so on. The important aspect here is the management of a particular resource, facility, area/region and so forth, so the statement would generally include the management activity. Table 2 shows some possible supports for decision-making from the GIS database. These could be used as references for defining education.

It is also important to specify the level or detail of the GIS database, which indicates the scale or level of the data contents of the database. A database designed for the micro-level would require far more details than one designed for macro-level applications. Table 3 illustrates the relationship between level and applications, which could be used as a guideline. In most of the cases the level of detail is implicit in the statement of the end use and is characterized by the data sets used.

Generally, the GIS database would be a set of natural resource information systems organized in two hierarchical levels, as follows:

(a) National-local level, catering to the national needs of spatial data and resource management needs for locale-specific sustainable development planning and implementation. Each member country could initiate activities towards establishing the national-local GIS databases as per their

Table 2. Possible GIS databases to support decision-making

Theme area	Possible support
Global monitoring	Global vegetation map from NOAA-GVI data of remaining forest areas. Global land cover classification by NOAA-GVI data for thirteen land-cover types. Estimation of supportable population by crop production.
Agricultural development	Environmental and agricultural development assessment. Habitability analysis for major urban centres. Satellite surveillance of agricultural drought conditions. Land evaluation of analysis for agricultural purposes; qualitative model for land suitability.
Deforestation	Remote sensing of shifting cultivation. Change detection of vegetated areas. Correlation between population density and forest cover in Asia and the Pacific region. Inventory of wastelands. Analysis of deforestation and associated environmental hazards.
Vegetation cover	Mapping percentage of vegetation cover for the management of land degradation. Monitoring vegetation health using NOAA-AVHRR data; regional vegetation conditions. Depletion of vegetation because of urbanization. Distribution of forest zones and forest mapping by means of remote sensing images and DTM. Monitoring of re-vegetation areas.
Natural resources	Crop acreage and production estimation; wasteland mapping; land use mapping for agro-climatic zone planning, forest mapping; soil resources mapping; groundwater potential mapping; assessment of damage due to floods and drought; geological and mineral exploration; snow-melt run-off forecasting; infrastructural development. Mapping of soil erosion and *jhum* lands. Characterization of environmental changes.
Desertification	Desertification prediction relative to the increase of shifting sandy land. Changes in vegetation and sand dunes. Monitoring and assessment of desertification in India.
Mapping from space	Topographic mapping. Bathymetric mapping. Weather charts.
Land use	Determining land-surface change in semi-arid environments. Modelling of spatial land-use changes. Land-use change analysis.
Disaster monitoring	Forest fire and post-conflagration change. Flood risk mapping. Cyclone damage assessment and movement.
Mangrove	Study of the changes of mangrove forests. Monitoring of the ecological status of the mangroves forests.
Ocean resources	Monitoring ocean productivity. General inventory of aquaculture site potential.
Snow and ice	Glacial and snow cover monitoring; snow-melt run-off forecast. Ice front fluctuations of glaciers. Snow mass climatology and distribution of snow mass.
Geology	Mapping of geological structures for ore-bearing areas; forecast for mineral resources. Hydro-geomorphological mapping for groundwater prospects. Erosional soil loss predictions.
Planning	Administrative planning for development of areas. Integrated land and water resource management.

Table 3. Relationship between GIS database detail and applications

Application	Extent	Spatial	Non-spatial
Micro-level	Micro-watersheds/cluster of villages/municipalities	1:25,000/1:10,000+	Village/plot level
Meso-level cities/urban	District/areas	1:50,000/1:25,000+	Village
Macro-level	State/region metropolitan region	1:250,000	Tehsil/panchayat
Global-level	Regions defined by countries, say Asia-Pacific	1:1,000,000	Tehsil

needs and resource management activities. ESCAP could recommend standards and guidelines for this level so that a standardization of formats and informative exchange would emerge;

(b) National-regional level, catering to spatial data needs at the regional and global level and oriented towards integrated resource management for preparing sustainable development strategies for implementation at the national-local level. ESCAP could take the lead role in establishing the regional-national level of the remote sensing information infrastructure (RSII) and coordinate with different countries.

The overall framework of these natural resource management activities will be encompassed by GIS databases at three levels:

(a) Regional level, where global assessment and programmes are formulated and guidelines developed for national strategies. These could be transnational agencies like ESCAP, FAO or UNDP, who could coordinate the global and regional-level GIS databases;

(b) National level, where the national policies are formulated and converted into plans and programs. These would be mainly the executive departments/agencies in each country and projects of the states;

(c) Local level, where the implementation of sustainable plans and programmes takes place, mainly the local field level organizations/departments and various projects within each country.

Some of the key issues that merit consideration for the GIS database design are discussed below:

(a) Spatial elements of the GIS database, which depends upon the end use and defines the spatial data sets that will populate the database. The spatial elements are applicationspecific and are mainly made of maps obtained from different sources;

(b) Non-spatial elements of the GIS database, which are the non-spatial data sets that would populate the GIS database. The actual definition of the non-spatial elements would depend upon the end use, and they are application specific;

(c) Impact of study area extent, defining the actual geographical area for which the GIS database is to be organized;

(d) Spatial framework, pertaining to the basic framework of the spatial data sets. Most of the spatial data sets follow the latitude-longitude graticule (as is given in the topomaps), so the spatial database needs to follow the standards of the topomaps. It is essential to adopt a standard registration procedure for the database. This is generally done by the use of registration points;

(e) Non-spatial data domain, specifying the levels of non-spatial data. The non-spatial data sets are available at different levels and it is essential to organize the non-spatial data at the lowest unit;

(f) Coordinate system for the database, which determines the way coordinates of spatial features are to be stored in the GIS database. Most GIS packages offer a range of coordinate systems, depending on what projection systems are adopted. The coordinate system for the GIS database needs to be in appropriate units that represent the geographic features in their true shapes and sizes;

(g) Spatial tile design, pertaining to the concept of a set of map tiles encompassing the total extent. Certain GIS packages allow for the organization of tiles, which facilitates systematic data entry on a tile-by-tile basis, as well as the horizontal organization of spatial data;

(h) Defining attribute data dictionary: the data dictionary is an organized collection of attribute data records containing information on the feature attribute codes and names used for the spatial database;

(i) Spatial data normalization is akin to the normalization of relations and pertains to finding the simplest structure of the spatial data and identifying the dependency between spatial elements. Normalization avoids general information and reduces redundancy.

O. Conclusions

The issue of designing and organizing a GIS database has to be considered in its entirety and needs a conceptual understanding of different disciplines -- cartography and map-making, geography, GIS, databases and more. In this report, an overview of the GIS package and its concepts has been provided. Different commercial GIS packages have also been covered here, to provide a survey of GIS packages.

Design procedures that may be adopted for organizing GIS databases, and the organizational issues, have been addressed. Based upon experience in organizing GIS databases for different projects, certain design guidelines have been framed which could be used as references for organizing GIS databases. The issue of updating the database and the linkage aspect of the GIS database to other databases has also been addressed.

PART THREE

SUSTAINABLE RURAL DEVELOPMENT PLANNING

INTEGRATION OF REMOTE SENSING AND GIS FOR SUSTAINABLE RURAL DEVELOPMENT

Y.V.N. Krishna Murthy and K. Radhakrishnan***

A. Introduction

In several parts of the world, misuse and mismanagement of land and water resources have set in degradation processes through loss of soil productivity and environmental quality. These are manifest in accelerated soil erosion, silting of storage reservoirs, rising stream beds, frequent floods, waterlogging, salinization and desertification. Associated with this problem, competing demands on land resources are gradually leading to land scarcity.

With the exponential growth of population and the goals of seeking self-sufficiency in food and fiber production, the resource base in the developing countries is slowly being stripped. Agricultural production in most of the developing countries ranges between a dismal 0.5 tons per hectare (t/ha) to 2.5 t/ha, as against the world average of 2.6 t/ha and over 4.5 t/ha in developed nations. With the projected increase in the population from the present 5.7 billion to over 8.5 billion by the year 2025, the situation is bound to be explosive, particularly considering the availability of only 0.17 ha per capita of land for agricultural production in the developing countries as opposed to over 0.6 ha per capita in the developed countries.

Agenda-21 of the Rio Earth Summit identified the necessity of integration of environmental and developmental issues for government decision-making on economic, social, fiscal, energy, agricultural, transportation, trade and other policies, with adequate public participation. It also emphasized that the increasing human demand for land, a finite resource, and for the natural resources it supports, is creating competition and conflict that results in land degradation, so an integrated approach to land use planning must examine all needs so that the most efficient trade-offs can be made. It also addresses integrated planning and management of all types of water resources through rational development plans that encompass multiple uses, including water supply and sanitation, agriculture, industry, urban development, hydropower generation, inland fisheries, transport and recreation, while conserving water and minimizing waste.

In order to optimize and sustain outputs from primary systems to meet the growing demands of rising population, developmental planning with an integrated approach has been accepted the world over. This approach helps attain optimal management and better utilization of natural resources, towards improving living conditions of the people.

1. Concept of integrated development

Integrated development is a comprehensive action programme aimed at optimal realization of resource potential in the light of physical, economical, social and other developmental goals. Such an endeavor entails harmonious development of land, water, vegetation and other resources of an area in a sustainable manner, so that the changes proposed to meet the needs of the development are brought about without diminishing the potential for their future use.

* Regional Remote Sensing Service Centre, Nagpur, India.

** NNRMS-RRSSC, ISRO HQ, Bangalore, India.

B. Integrated planning approach and goals

1. Approach

A practical approach in planning, directed at preservation, conservation, development, management and exploitation of the natural resources of the region for the benefit of the people has to operate within the framework of:

(a) Physical and biological attributes;
(b) Socio-economic conditions;
(c) Institutional constraints.

Physical and biological attributes comprise baseline data on geomorphology, geology, soils, hydrogeology, hydrology, climate, demography, and plant, animal and other biological resources. Socio-economic conditions relate to information on basic needs of the people, input-output relationships, marketing and transportation arrangements, developmental incentives and facilities, such as technologies, equipment, labour, material, energy and power. Institutional constraints related to laws, regulations and ordinances, government policies and priorities, political acceptability, accepted customs, beliefs and attitudes of the people, and administrative support.

The separate inventories of the physical and biological attributes, socio-economic conditions and institutional constraints are integrated using GIS. This would provide the physico-socio-economic profile of the region/watershed and permit suitable development models in the different sectors of the economy and production. The system as a whole would thus be operated upon to develop appropriate alternatives of conservation-production programmes commensurate with the production potential. For example, a particular land unit, in conjunction with water, could be profitably and sustainably exploited for agriculture or forestry or other uses, consistent with its land capability evaluation so as to maintain ecological balance, provide protection against soil loss and land degradation and deliver a clean and controlled flow of water. Likewise, based on hydrologic groupings of soils, appropriate water management programmes would be identified not only to improve water regime and water use, but also to protect the land against waterlogging and salinization.

2. Prelude to inventory

Each region/watershed has its own problems, potentials and development needs, defined by a set of geophysical, climatic, demographic and socio-economic factors. It is necessary to have a broad idea about these aspects prior to information generation, primarily for identifying the planning goals and deciding the abstraction level of thematic mapping in each case. The information should relate to specific needs of the people, overall potential of the area, general problems of the area, availability of information, gaps in the information, and acceptability of the information.

3. Planning goals

Based on the preparatory information, the following broad planning goals have been identified:

(a) Provide for basic needs of the people -- water, fuel, food and fodder;
(b) Develop and optimize primary production systems and practices -- agriculture, forests, grasslands, fruit and other economic plantations;
(c) Control soil erosion/land degradation and reclain degraded lands;
(d) Soil conservation, sediment control and runoff moderation;
(e) Optimize production minerals with proper plans for rehabilitation of mined areas;
(f) Restore wastelands to their production potentials consistent with land capability classification;
(g) Develop and manage water resources -- surface and ground;
(h) Optimize irrigation and management of irrigated agriculture;
(i) Promote animal husbandry, dairy development and poultry;
(j) Industrial growth, environmental security and improvement of socio-economic conditions.

C. Databases

Developmental planning is a complex process of decision-making, based on information about the status of resources, socio-economic conditions and institutional constraints. Reliability of the databases, both spatial and non-spatial, is therefore crucial to the success of development planning. Equally important is the timely inflow of information to serve planning needs. Remote sensing technology, which meets both the requirements of reliability and speed, is an ideal tool for generating spatial information bases. The databases are meant to serve an efficient system of information gathering, compiling, classification, transformation, storage, retrieval, synthesis/analysis and presentation.

1. Resource database: spatial

Consistent with the region/watershed-level planning requirement, thematic maps shall be generated. Both the digital and visual techniques shall be followed interactively. Special techniques of stratification, layered approach and composition, aggregation, and refinements shall be adopted wherever necessary to improve the quality of mapping. The primary thematic maps needed are shown in table 1.

Table 1. Primary thematic maps

Theme map	Source	Remarks
Land use/land cover	IRS LISS-II and III (multi-date), ground truth	Digital classification using stratified approach. Refinement using human logic.
Soil	IRS LISS-II and III and available maps, profiling	Digital enhancements and visual interpretation. Digital classification using stratified approach.
Geology	IRS LISS-II and III and field data existing geological maps	Digital enhancement and mapping. Digitization.
Geomorphology	IRS LISS-II and III and topographical maps	Digital enhancement and mapping. Digitization.
Drainage and sub-watershed boundaries	Topographical maps and IRS LISS-III data	Digitization.
Digital elevation	Survey topomaps and IRS-1C PAN (stereo)	Digitization of contours/DEM derivation from stereo image.
Transport network	Topographical maps and IRS-1C PAN data	Digitization of features. Updating of images.
Village boundaries	Census book/maps	Digitization.

2. Spatially derived databases: spatial

Basic maps are used to produce utilitarian types of maps to serve planning decisions. They are derived, in some cases, by direct translation of a single thematic maps, and in others by combination of two or more thematic maps or chosen parameters of the different themes (table 2).

3. Attribute database

As mentioned earlier, socio-economic conditions and institutional constraints greatly influence the development programmes. Voluminous information on these aspects exists at various sources and different levels, which needs to be ascertained and quantified. These are demographic, inputs status, facilities, sociologic, financial, policies and priorities.

4. Ground truth collection

The scientists responsible for compilation of particular thematic maps shall collect ground truth information from the predetermined sites based on image manifestations on satellite data. The ground truth sites shall be distributed all over the region/watershed area in such a way that they collectively

Table 2. Maps derived from thematic maps

Derived map	Theme map	Remarks
Slope	Topographic map/IRS PAN stereo data	Derived from DEM.
Land capability	Soil, slope, climate	Digital aggregation.
Land irrigability	Soil, slope, land-form, groundwater, depth, Exch. SAR, EC	Digital aggregation.
Groundwater potential	Geology, geomorphology, borewell, litho-log and yield data	Intersecting of theme layers and abstraction of point database.
Run-off potential	Slope, soil map, land use, rainfall	SCS model through integration of layers.
Run-off depth	Slope, soil map, land use, rainfall, sub-watershed map	SCS model through integration of layers.
Peak run-off rate, sub-watershed-wise	Slope, soil map, land use, rainfall, sub-watershed map	SCS model through integration of layers.
Run-off volume, sub-watershed-wise	Slope, soil map, land use, rainfall, sub-watershed map	SCS model through integration of layers.

cover all the variations in the region/watershed. About 10-15 per cent of the total area shall thus be tested on the ground. Satellite images and topographical maps shall be used for locating the sites and plotting the information. In the case of soils, auger bores and mini-pits shall be examined in addition to a limited number of soil profiles at suitable sites. Post-classification groundwork shall be carried out to verify the mapping units and their boundaries for validation of maps.

D. Data integration and development alternatives

1. Mechanism for integration and manipulation

The integration of the various thematic maps and attribute data, and further manipulation/analysis for identifying alternatives for development, shall be carried out using a state-of-the-art geographic information system. Expertise is available in accomplishing the task on any software, either PC-based or on a workstation.

The digitally classified outputs corresponding to geology, geomorphology, soils, land use and their derivatives shall be feature-coded and stored in the map information system. These individual maps from corresponding map files shall be integrated to arrive at "composite mapping units" (CMUs). The socio-economic, institutional and other statistical data would be entered into the attribute data-base. The decision criteria would be structured within the framework of resource potentials and other determinants to evolve a pragmatic model.

2. Composite mapping unit

A composite mapping unit is a three-dimensional landscape unit homogeneous with respect to characteristics and qualities of land, water and vegetation and separated from other dissimilar units by distinct boundaries. The CMU characteristics imply physical parameters of the component resources of a biophysical domain, whereas qualities are suggestive of their potential for specific uses under the defined sets of conditions. Based on the interaction among the basic resources of land, water and vegetation which form the major components of a primary production system, useful inferences can also be drawn about their predicted behaviour in meeting the various planning goals.

3. Current status of resource utilization and management

To begin with, all collateral data derived from the District Census Handbook, Agricultural Census Report, meteorological tables and from other relevant sources, in conjunction with land-use maps, are assembled and collated to assess:

— Man-land ratios (man: total land; man: arable land; man: forest land);
— Present land use;
— Present water use;
— Local problems;
— Agricultural production;
— Employment status;
— Marketing, storage;
— Input situation.

Based on the above data, surpluses and deficits are computed in each area at the present level of demand and supply. Projections for the future demands are then computed considering the current population growth rate. This would lead to identification of thrust areas for the different *taluks* in the district. It is unlikely that the resource potentials of a particular taluk, even with an ideal plan, would meet all the variegated needs of the people, but it would help to ensure that the CMUs have been collectively manipulated to their optimal use without losing sight of the local problems and needs. It may be mentioned that self-sufficiency can hardly be a realistic goal at the planning level of the region/watershed.

The areas under agriculture, forests, grasslands and plantations suffering from soil erosion and other forms of degradation reflect the extent of mismanaged lands, for which integrated soil and water conservation measures are indicated. The wastelands are the locked-up production potential and call for appropriate reclamation measures. For a lasting solution to these problem, both the reclamation and the soil and water conservation problems have to be identified on the basis of characteristics of the associated soils and input situations. Land-forms and slope are other important determinants for ameliorative and preventive measures.

A land capability model based on soil, slope and meteorologic factors gives optimum land use potential. A mismatch between existing and potential land use implies land use revision. A land irrigation model helps to assess the suitability of land for irrigated agriculture and enables predictions of its behaviour under the defined management level. Hydrologic groupings are indicative of infiltration/surface flow for soil, based entirely on its inherent characteristics. Integration of hydrologic groupings, vegetation cover complex, land-forms and slope permit run-off estimation from a given land system unit.

Integration of geological, geomorphological, hydrogeological and land use data with geophysical investigations gives groundwater potential. This, coupled with surface water potential, when matched against tapped water resources, helps in estimating unrealized water potential to meet the primary demands of irrigation, industries, drinking water and others. These are only some of the examples of manipulation of CMUs.

4. Development of land/water use alternatives

A pragmatic development model has to provide a number of alternatives with respect to each of the different landscape units or CMUs. Primarily, this is because our understanding about the interactions between the different parameters of the complex social, physical and natural system is too limited to suggest a single best course of action. Furthermore, the input situation and socio-economic conditions assumed at the planning stage for a particular action plan may not remain valid at the time of implementation, thereby rendering the plan ineffective. There is also a risk factor involved in a single action plan that is based almost entirely on the "assumed best" in alliance with goals and objectives of the planning, although such an action may not necessarily agree with the individual landholder. Alternatives provide a choice to the landowners to select from amongst several recommended action plans and thus indirectly help to overcome their resistance to something that at least some of them may feel is being thrust upon them.

Alternatives are developed within the framework of optimal land use. On applications of the land capability classification (LCC) model, an optimal, broad land-use category is derived from the CMU. This, when matched with present land use (indicated by CMU), helps decisions on broad land use revision matching the LCC. The need for specific programmes of development, conservation and management is assessed by application of the runoff potential classification (RPC) model, land irrigability classification model (LIC) and productivity index value (PIV). PIV, when rated for specific crop against the present level of production per unit area, helps assess the gap that has to be bridged by adoption of recommended practices. PIV in monetary terms helps in revision of cropping patterns for introduction of climatically adapted high-value crops. Irrigation and other input conditions, as well as socio-economic priorities of the area specific to the CMU, are never lost sight of in formulating land and water use programmes leading to options of land and water utilization types and practices.

Finally, the alternatives should be developed to cover:

- All the systems to be planned;
- All the sectors of the economy;
- All the sectors of the community.

5. Plan implementation and monitoring

The survey boundaries (parcels) are the smallest units and are essential in identifying government and private lands, for plan implementation and subsequent monitoring of the impact on the environment. The cadastral maps for each village are available on 1:5,000 scale. These maps depict the survey boundaries with the survey numbers, cultural features like the transport network, and natural features like drainage. The cadastral boundaries overlaid on the satellite images help in prioritizing the plan implementation and monitoring. IRS-1C PAN data has been effectively used for cadastral boundary overlays and updating for speedy implementation of the development programmes.

An increasingly useful application of geographic and land information systems in cadastral systems is to provide up-to-date records of land tenure, land values and land use in both textual and graphic formats. In such a system, the parcel is the principal unit around which the collection, storage, and retrieval of information operate. The information contained in a cadastral system makes it possible to identify the extent and level of investment, development and management in land and (assuming the quality of information in the cadastral system is adequate) to make effective plans for the future.

The satellite data from IRS LISS-II and LISS-III, once in every cropping season, should be used to estimate the following changes:

- Increase of area under vegetation;
- Increase in cropping intensity;
- Increase of irrigated area;
- Soil moisture retention;
- Reduction of soil erosion.

This approach and methodology has been operationalized through a package, Geo-LAWNS, designed and developed by Department of Space for integrated land and water resource planning for rural development.

A SYNERGISTIC APPROACH TO SUSTAINABLE RURAL DEVELOPMENT USING SATELLITE REMOTE SENSING AND CONVENTIONAL DATA

*S.K. Subramanian**

A. Introduction

The evolution of mankind and exploitation of the natural resources are highly correlated. Human beings are the single largest force creating imbalances in the ecological systems of the earth in the name of development. The regard, so far, for carrying capacity of the land, resilience of the ecosystem or ecologically safe development has been negligible. Fortunately, changes in the level of population during the last century have compelled planners, worldwide, to realize that our natural resources are not going to be there forever and that, unless population growth is checked, we are moving fast from an era of plenty to scarcity. Another thought currently receiving attention is conservation and development of natural resources themselves so that they can be used on a sustainable basis. *Resource sustainability* means that we not only have sufficient resources for all our requirements but also leave behind enough for our children and children of our children as an ecological heritage. The concept of sustainable development has become very popular since the Rio Conference in 1992 because of its paramount importance. Sustainable resource development can be defined as development and management of natural resources to meet the immediate needs of the present population and the requirements of future generations without, in any way, endangering the productive potential of the resources and the environment. Sustainable development of natural resources relies on maintaining the fragile balance between productivity and conservation functions through precise identification and systematic monitoring of problem areas in various resources and developmental sectors.

B. Sustainable development

The developmental activities of the backward regions in India have been mostly based on the criterion of economic growth. While there is no option but to produce more food and improve necessary socio-economic conditions, it is also important to recognize that today's economic progress should not be at the expense of tomorrow's developmental prospects. Hence, the resource-based development plan is a must for sustainable development. Effective use of space-based remote sensing data, suitably blended with socio-economic data, helps not only in arriving at locale-specific prescriptions to achieve sustainable development of a region but also in monitoring the impact of developmental activities taken up under various developmental schemes, especially in enhancing the rural economy.

By the year 2020, the Earth's population may reach 8 billion, out of which the developing countries alone contribute 83 per cent of the population. So there is a need to increase the food production from the present 1,800 million tons to 3,000 million tons. The present rate of urban and industrial development also eats away at the productive farm land -- so the increase in production will be restricted to even less farmland than is available today. Therefore, there is a need to identify sustainable techniques, which should address the agricultural needs in terms of quantity, quality and cost of satisfying market requirements for food and fuel for humans and fodder for animals. In addition, the areas for such activities have to be earmarked. In this aspect, wastelands and marginal lands are important for transforming into productive land. Such lands can be identified by using space technology, and the suitability of the land for crop, pasture or fuelwood can be determined by incorporating various remote sensing methods.

* National Remote Sensing Agency, Hyderabad, India.

C. Space technology and sustainability

The Department of Space has taken up the task of identifying and estimating the areal extent of the wasteland nationwide and estimated the total area to be 755,153 sq km. The marginal lands can also be turned into useable lands by bringing the employment factor into consideration, especially for the rural poor, most of whom are living under the poverty line. These areas have to be used for fuelwood and fodder production. This will not only take care of the ecological system but also provide employment and development on a sustainable basis. It is clear that space technology can help us take care of the ecological system and help us provide employment and development on a sustainable basis. Space technology not only addresses the task of identifying the suitability of cropping patterns but also helps in monitoring the development of specific locales. The present trend in developing the rural economy is based on the watershed development, wherein space technology contributes in mapping the natural resources and identifying the locale-specific action plan. As most people are aware, rural India is a land of small holdings -- about 70 per cent of the population depends on agriculture for jobs and income. Under such socio-economic conditions any developmental plan will have to be very carefully formulated. It has to be based on resource availability, need, economy, problems and acceptability of people without endangering the ecosystem, and such a plan for development has to be on a sustainable basis.

It is well known that remotely sensed data from satellites has the advantage of repetitive, synoptic views of large areas in a multispectral mode. Since 1972, the Landsat series of satellites has provided spatial resolutions starting from 80 m and down to 30 m, and later SPOT satellites provided resolution in panchromatic mode down to 10 m. The scales of mapping improved with resolution, making it possible to map up to 1:50,000 scale. Indian Remote Sensing satellite series started in 1988 with IRS-1A and the later IRS-1B, followed by IRS-P2, which provides resolutions that make it possible to map up to 1:50,000 scale.

Any study addressing the above issues needs an elaborate database on natural resources, physical/terrain parameters, climate, contemporary technology and the socio-economic profile of the study area. Analysis of the data and integration of the same with contemporary technology will help in arriving at an action plan for sustainable development. The conceptual framework of the study taken up by the Department of Space under the Integrated Mission for Sustainable Development (IMSD) is shown in figure 1.

D. Rural development applications: a case study

The objective of this study is to develop and demonstrate the methods for watershed-level natural resource development planning using remotely sensed data and other information on climatic and socio-economic conditions for sustainable development. Out of 174 districts/watersheds of under-developed regions in India, where the IMSD study is undertaken, a part of the Upper Hatni watershed in Jhabua District, Madhya Pradesh, has been chosen for this study.

The upper Hatni watershed lies in the eastern part of the Bhabhra block (22°32'-22°37'N latitude and 74°17'-74°25'E longitude) in Jhabua District, M.P. (figure 2). A sub-watershed measuring 9,085 ha forming the upper part of the Hatni watershed was selected. The average elevation in the area is 340 m above mean sea level, comprising hills and undulating terrain. The area has steep to moderate slopes varying from 1-5 per cent in plain areas and 15-35 per cent near valley sections. The surface run-off is high. Geologically, the area comprises Archaean granite, quartzite and phyllite, Bagh sandstone, basaltic flows and alluvial deposits. Geomorphologically, this area can be divided into valley fills, basalt lava plains, pediment, buried pediment and denudational hills. Soils in the area are loamy skeletal on hills, shallow fine loamy in the foot hills and fine loamy fluventic on lower sections. The extreme northern part is covered by clayey skeletal soils.

The area is, for the most part of the year, unsuitable for permanent agriculture. Improper land-use practices, such as agriculture on steep slopes, have resulted in accelerated erosion of the soil cover, so over time the hills are laid bare. The area has chronic water scarcity. This, together with poor soil, results in frequent crop failures or low crop yields. Sufficient fodder is also not available for the large cattle population. Extensive grazing has been detrimental to the growth of the vegetation. The degradation of the forest area is alarming. The area needs proper land and water management measures to check soil erosion, improve groundwater and surface water situations and to arrest the forest degradation. The area under kharif crops, e.g. paddy, maize, cotton and pulses, is 87 per cent. Only 8 per cent is used for both kharif (July-October) and rabi (November-March) crops. This is because of the lack of adequate water in storage dams, tanks, wells and in the soil profile. The average rainfall is about 600 mm, occurring during the second week of June to the middle of October, mainly from the south-west monsoon.

Upper Hatni watershed is composed of 24 villages and a population of 18,130 (1981 census), of which nearly 97 per cent are scheduled tribes (hill tribes). The majority of the people practice agriculture and depend on their marginal lands for sustenance. Animal husbandry is another favoured occupation and the area is known for its unusually large animal population. This area is also visited annually, like most other regions of Madhya Pradesh, by sheep and goat-rearing nomads of Gujarat and Rajasthan, which increases the already tremendous pressure on grazing land. The majority of the population is illiterate and backward.

IRS-1B LISS-II standard false colour images at 1:50,000 scale of kharif, rabi and summer seasons were used for this study. Ancillary data like Survey of India topographic sheets, census reports and the information available from the state line departments were also used for this purpose.

The whole task was divided into two parts: (a) natural resource-related information generation and (b) information integration and analysis.

1. Information generation

The information on geomorphology, existing land use, soils, surface water and groundwater was derived through visual interpretation of false colour imagery at the 1:50,000 scale. A common base map was prepared and theme details of each resource were transferred onto it to maintain consistency and positional accuracy with respect to base details. Interpretations were made by individual subject specialists. An effort was made to achieve more than 90 per cent mapping accuracy, keeping in mind the importance of preliminary interpretation, field checking, corrections, rechecking and final corrections by a joint team of scientists from the National Remote Sensing Agency (NRSA), Hyderabad, and the Madhya Pradesh State Remote Sensing Application Centre, Bhopal, in cooperation with Jhabua District administrative authorities. The information on rainfall, temperature and socio-economic conditions was gathered from district records, while transport network, slope and drainage were drawn basically from Survey of India topographic sheets and augmented, wherever necessary, by false colour imagery.

2. Information integration and analysis

The second but more important part of the study was the integration of all information generated from various sources to suggest suitable action plans for sustainable development of the watershed. In this process, help came in the form of the field research work for similar areas that had been carried out by various Indian Council of Agricultural Research (ICAR) centres and agricultural universities. The contemporary technology in the area of resources management and development was also taken into consideration. Also, discussions were held with district development authorities and farmers. The process of integration adopted in this study was manual. The same, however, could also be accomplished using a geographic information system (GIS). Each parcel of land was evaluated with respect to land-forms, soil, slope, groundwater, land cover, surface water, rainfall and other factors to find out its suitability for a particular land use. Before suggesting any land use, due consideration was given to the demand and supply situation of the human and animal populations. The preliminary recommendations were further subjected to various ecological and environmental considerations to decide whether they were environment-friendly and ecologically safe. The basic rules formulated for the information integration are shown in table I. The recommended land-use practices are in the form of an action plan map (at 1:50,000 scale) showing areas suggested for silvipasture, fodder and fuelwood plantations, agroforestry, agro-horticulture, forestry, agriculture and rainwater harvesting structures (figure 3).

3. Packages of recommended activities

This study resulted in packages of recommended activities that were considered most suitable for the upper Hatni watershed. Considering the terrain conditions, water availability situation and the type of occupation, most acceptable to the local people, the following farming systems have been suggested.

(a) Silvipasture

Silvipasture is proposed to be developed in about 1,380 ha of land. It is essentially a system of growing trees and grasses together on the same plot of land to maximize fodder and fuel wood production on dry, skeletal, sloping lands. The trees should be fast growing, be resistant to drought and have fodder and fuelwood value. The tree species recommended are shisham (*Dalberzia sissoo*), babul (*Acacia nilotica*), mahua (*Madhuca latifolia*), neem (*Azadirachta indica*), bamboo (*Dendrocalamus* sp) and khair (*Acacia catechu*), among others. The grass/legume components selected are based on their ability to propagate prolifically and tolerate the drought. *Dicanthium, Themedae, Pennisetum* sp and *Cenchrus ciliaris* among grasses and *Stylosanthes hamata* among legumes are promising species for the area. These species in mixture would not only provide fodder and fuelwood but also enrich the soil with nitrogen, making the same land suitable for a higher order of utility. Fencing of the fields would be required. This can also be done by digging cattle prevention trenches along field boundaries.

(b) Agroforestry

Agroforestry is proposed to be developed in about 5,135 ha of land. Agroforestry is the modern name for an age-old practice of growing trees along with the crops in agricultural areas. Particular care is taken to select trees and crops that mutually benefit from the association. The tree species having fodder, fuelwood and timber value such as *Lucaena leucocephala*, *Pithecolobium* sp and others are suggested under this activity. The crop species are millet, sorghum, maize, pigeon pea, horse gram, green gram and black gram. Live fencing with agave and euphorbia has to be done in order to protect the land from grazing animals.

(c) Agro-horticulture

It is proposed to develop agro-horticulture in about 795 ha. The fruit trees recommended are *Ziziphus mauritiana*, anola (*Emblica officinalis),* tamarind (*Tamarindus indica*), mango, jamun (*Syzygium cuminii*), bel (*Aegle marmelos*) and sitaphal (*Anona squamosa*). Crop species suggested are pigeon pea, black gram, millet, sorghum, soybean, green gram and cluster bean. It is already well known that benefits of growing fruit trees and crops are more remunerative than either one of them alone, particularly in semi-arid and arid regions. Live fencing is recommended along field boundaries.

(d) Agriculture

About 1,700 ha of land has been identified where double crops can be grown. Valley areas could be suitable for paddy cultivation. Paddy in combination with maize can be grown in other areas. Paddy may be followed by castor during the rainy season and by gram in post-rainy season. Demonstration of improved agricultural practices and extension of necessary agricultural facilities are also proposed under this activity.

(e) Rainwater harvesting

Rainwater, if harvested appropriately, can mitigate the drought situation to a large extent. It is, therefore, suggested that rainwater harvesting structures like contour bunds, check dams, farm

Table 1. Basic rules for integration of the thematic information

Land-form	Soil class	Groundwater	Slope	Land cover
I. Silvipasture				
1. Denudational hill	Typic ustorthents Typic ustochrepts	Poor to nil	>8 per cent	LWWS
2. Denudational (marginally cropped) hill	Lithic/typic ustorthents	Poor to nil	>8 per cent	SCA
3. Dissected land, plateau	Typic ustorthents	Poor to moderate	>8 per cent	Gullied grazing land
4. Dissected (marginally cropped) plateau	Lithic/typic ustorthents	Poor to moderate	>8 per cent	SCA
5. Plateau grazing (basalt)	Lithic ustochrepts	–do–	5-8 per cent	LWWS, land
6. Linear ridge gullied	Typic ustorthents	Nil to poor	3-5 per cent	LWWS, land
7. Buried pediplain grazing	Udic ustochrepts	Poor to moderate	3-5 per cent	LWWS, land
8. Pediment	Typic ustorthents, Udic ustochrepts	–do–	3-5 per cent	–do–
9. P-I complex	Typic ustochrepts	–do–	3-5 per cent	–do–
10. Undulating pediplain	Lithic ustorthents, Typic ustorthents	–do–	3-5 per cent	–do–
II. Agroforestry				
1 Pediplain	Udic ustochrepts, Udic/typic ustorthents	Poor to moderate	3-5 per cent	Fallow
2. Dissected plateau	Typic/udic ustorthents	–do–	3-5 per cent	SCA, fallow
3. Plateau (basalt)	Typic/udic ustochrepts	–do–	3-5 per cent	–do–
4. Buried pediplain	Udic haplustalfs, Udic ustochrepts	Moderate to good	3-5 per cent	–do–
5. Pediment	Udic haplustalfs, Udic ustochrepts	–do–	3-5 per cent	–do–
6. P-I complex	Typic ustorthents Udic haplustalfs	–do–	Up to 3 per cent	–do–
7. Undulating pediplain	Typic ustorthents, Udic ustochrepts	Poor to moderate	1-3 per cent	–do–
III. Agro-horticulture				
1. Pediplain	Udic/typic haplustalfs, Fluventic ustochrepts, Typic chromusterts	Moderate to good	1-3 per cent	SCA
2. Dissected plateau	Typic chromusterts	–do–	1-3 per cent	SCA
3. Plateau (basalt)	Typic chromusterts	Poor to moderate	1-3 per cent	SCA
4. Buried pediplain	Typic chromusterts, Fluventic ustochrepts	Moderate to good	1-3 per cent	SCA, agricultural, plantation
5. Pediment	Fluventic ustochrepts, Typic ustorthents	Poor to moderate	1-3 per cent	SCA
6. Valley fill	Fluventic ustochrepts, Typic ustorthents	Good to excellent	1-3 per cent	SCA, agricultural, plantation

LWWS -- Land with or without scrub.
SCA -- Single-cropped area.

ponds and percolation/storage tanks should be constructed wherever possible, to arrest the free flow of rainwater. Vegetative bunding is another alternative. Bunding with subabul, sesbania and cassia is recommended for the area. A rise in the groundwater table after construction of check dams has been reported in the Vanjuvanka watershed of Anantapur District, Andhra Pradesh, under the IMSD programme (APSRAC, 1993). Such measures would provide life-saving water for agricultural fields, forests and horticultural plantations and drinking water for livestock and human beings.

(f) Animal husbandry and dairy development

The area has good potential for development of animal husbandry, which is also a natural choice, considering the inclination and expertise of the local people. Poultry and dairy development are two major activities proposed for this area. Development of veterinary services, infrastructure for collection and processing of milk and other livestock products, training and other activities have to be taken up simultaneously.

(g) Allied activities

It will be necessary to develop roads and markets for agricultural and horticultural products, and for milk and livestock products. The training and education of local people through demonstration schemes, audio-visuals, cultural shows and other activities should be taken up. Extension of health care facilities such as immunization and mother and child care should also be given due importance for simultaneous and proper development of human resources. The implementation of these activities is also being planned.

(h) Implementation

The results obtained from the IMSD studies have been implemented in the field with the involvement of user agencies and state government departments. The action schemes like afforestation, agroforestry, gully plugging, contour trenching and check dams for ground water recharge have been undertaken in 38 watersheds covering an area of 20,593 ha, costing Rs 69.95 lakh (6,995,000 rupees) in Jhabua District.

E. Information on cadastral level

With the launch of the IRS-1C satellite on 20 December 1995, a new dimension was been added to India's imaging capability. IRS-1C provides data at 23 m resolution in multispectral mode and 5.8 m in panchromatic mode. Therefore, IRS-1C satellite data can be used for the development of rural areas on a larger scale at the cadastral level. A study on this aspect was carried out by NRSA in part of a watershed in Dharmapuri District in Tamil Nadu and the results were also validated in the field (figure 4).

F. Conclusion

India's Integrated Mission for Sustainable Development, combining space remote sensing inputs on land and water resources with collateral socio-economic information, provides a holistic approach for achieving sustainable development to meet the growing needs of the increasing population. Such an integrated mission provides a unique planning approach, in contrast to the sector-based and beneficiary-oriented ad hoc approaches followed so far. Successful accomplishment of these goals, however, involves the concerted efforts of several central and state departments, experts in different fields, non-governmental organizations and, above all, the participation of local people. The experience gained from pilot experiments has conclusively demonstrated that implementation of action plans generated through IMSD, when integrated appropriately with the ongoing central and state schemes, will help in realizing the all-round development of the country at the village level. The unique methodology of using space technology inputs provides clear, unambiguous and quantitatively measurable parameters for assessing the effectiveness of the mission. Implementation of similar sustainable integrated development strategies is the only hope for developing adequate food and economic security on a sustainable basis in many poor developing countries.

Reference

APSRAC (1993). Integrated study to combat drought on sustainable basis through space applications: Vanjuvanka watershed, Ananatpur District, Andhra Pradesh. Andhra Pradesh State Remote Sensing Applications Centre, Hyderabad.

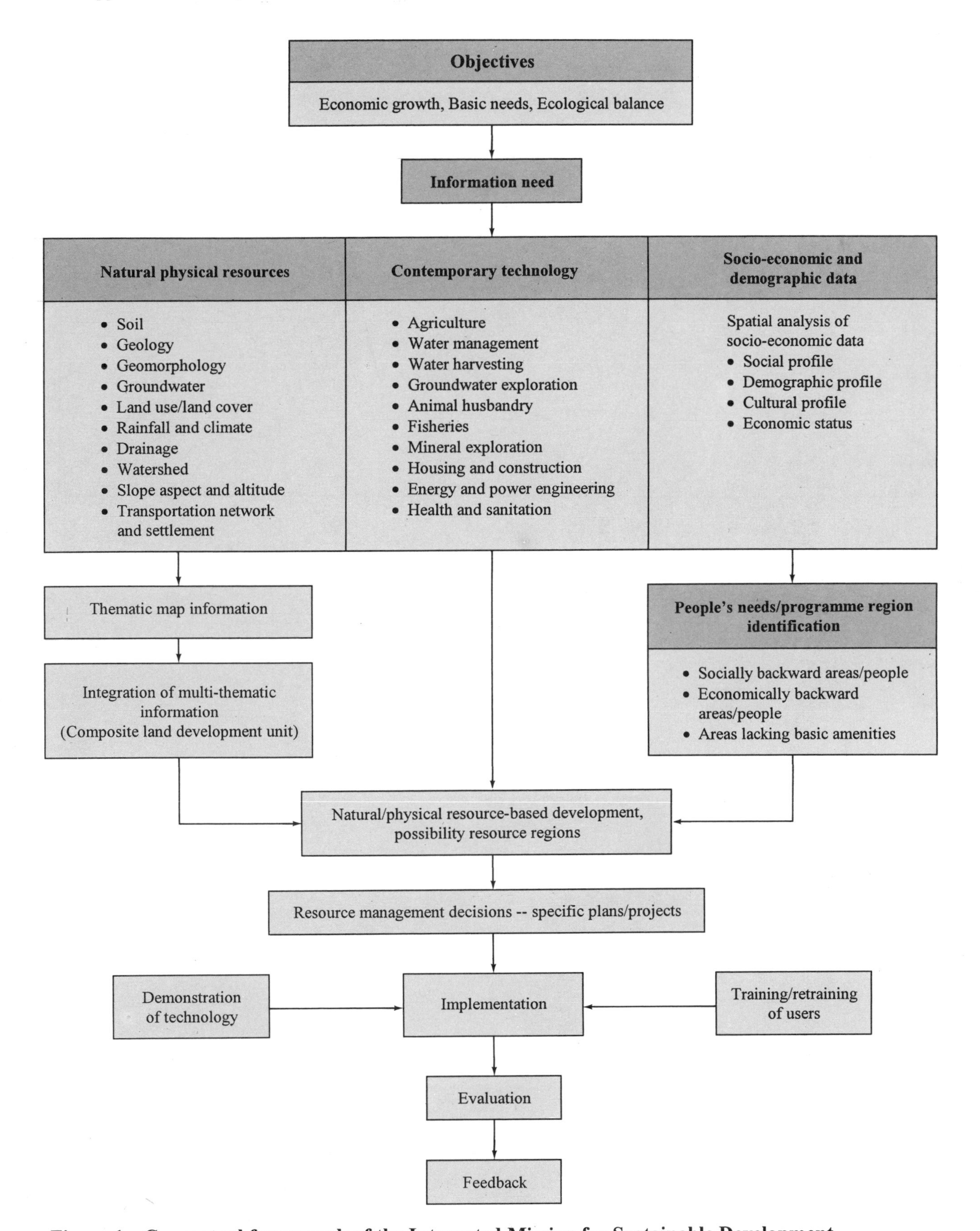

Figure 1. Conceptual framework of the Integrated Mission for Sustainable Development

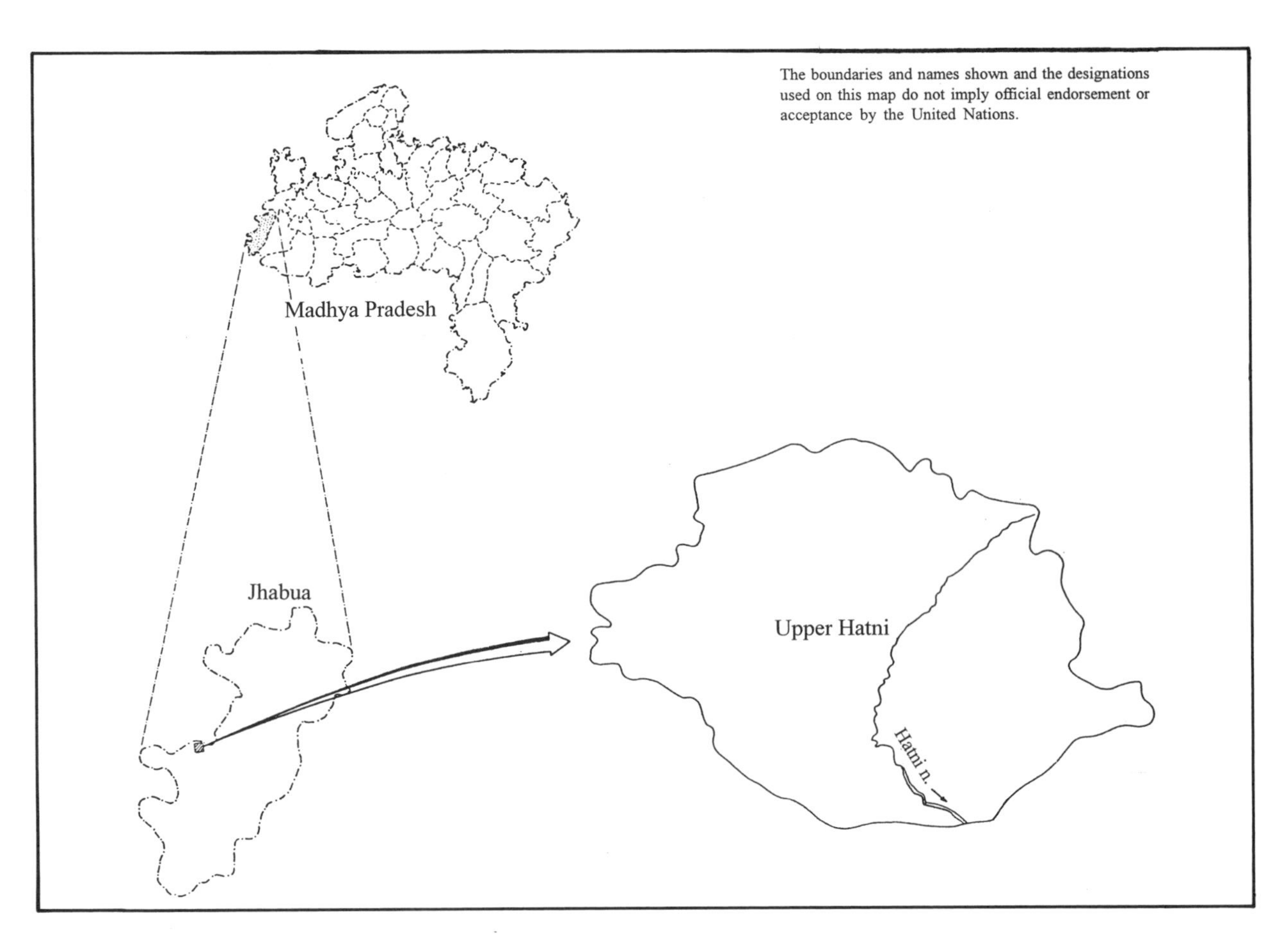

Figure 2. Location map of Upper Hatni watershed

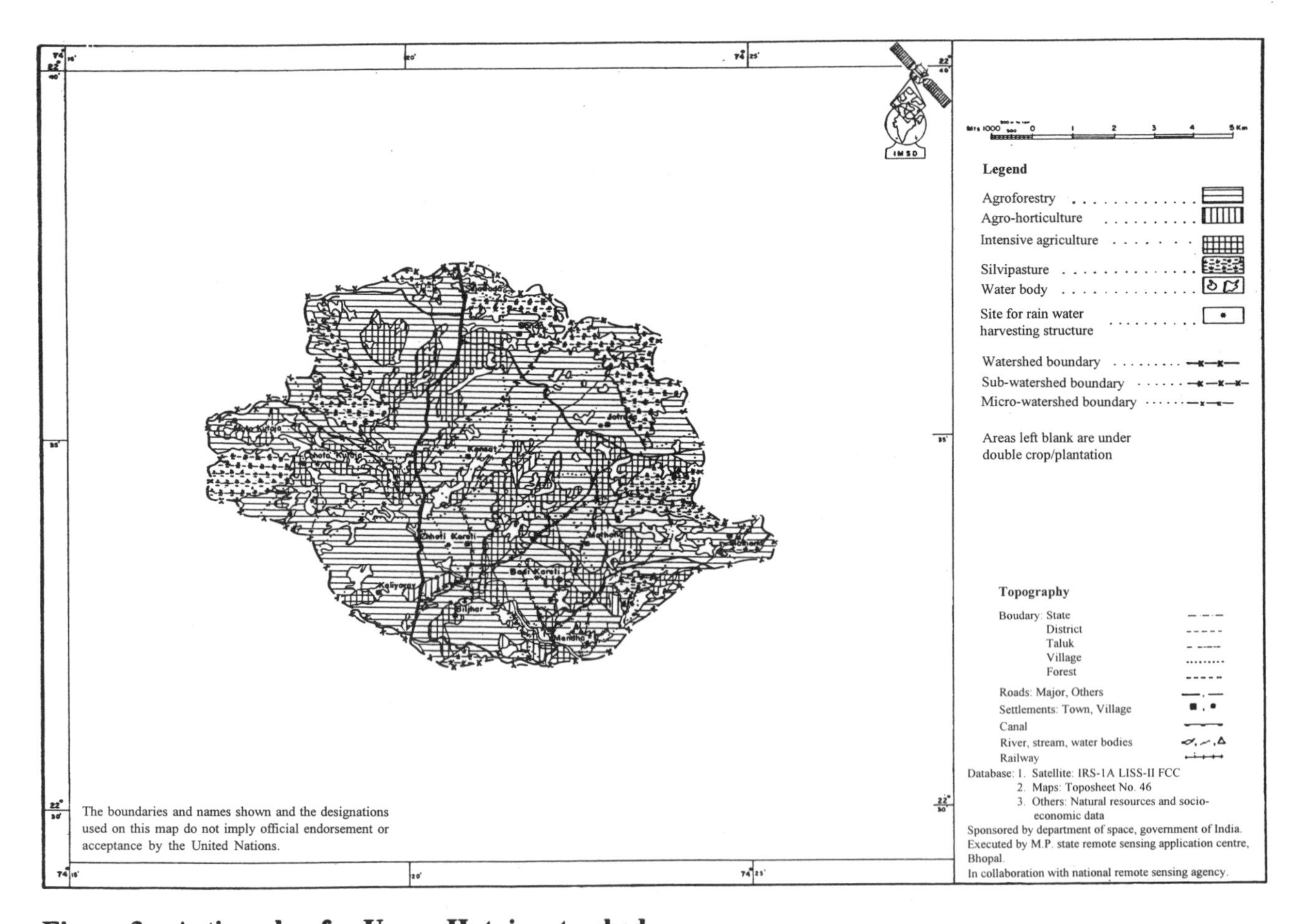

Figure 3. Action plan for Upper Hatni watershed

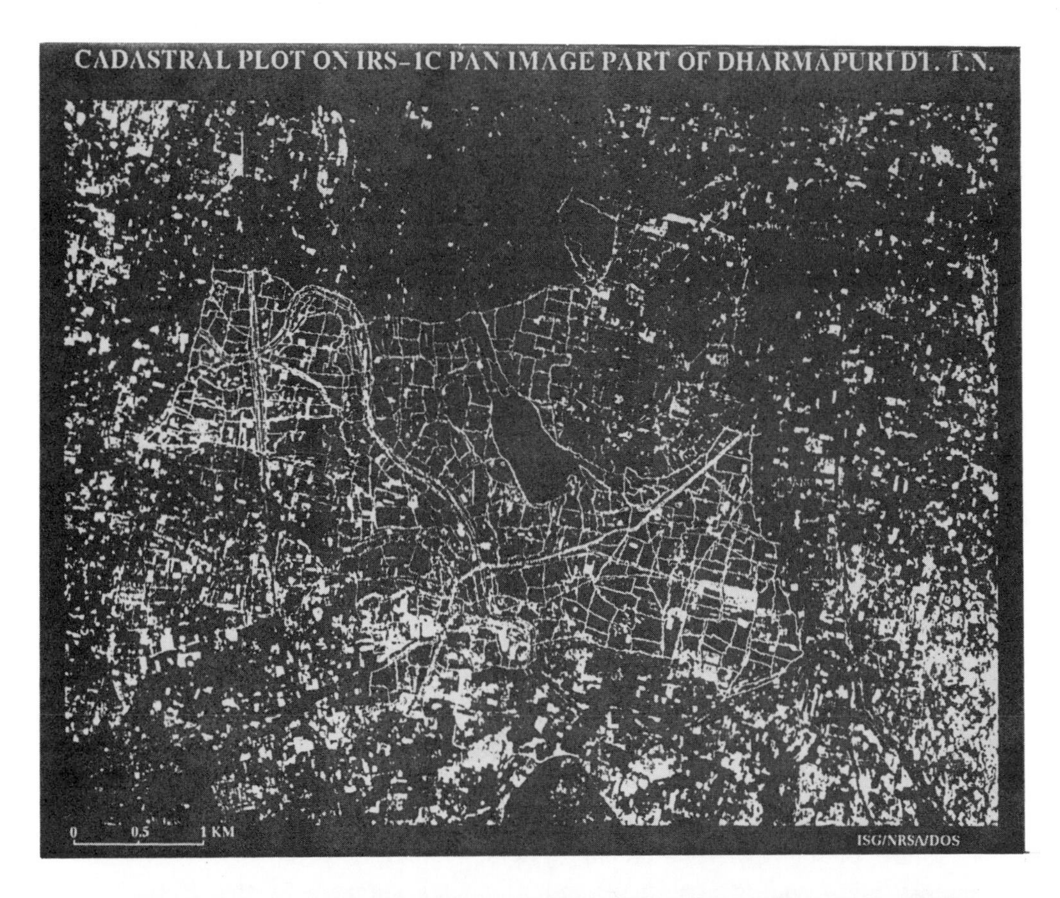

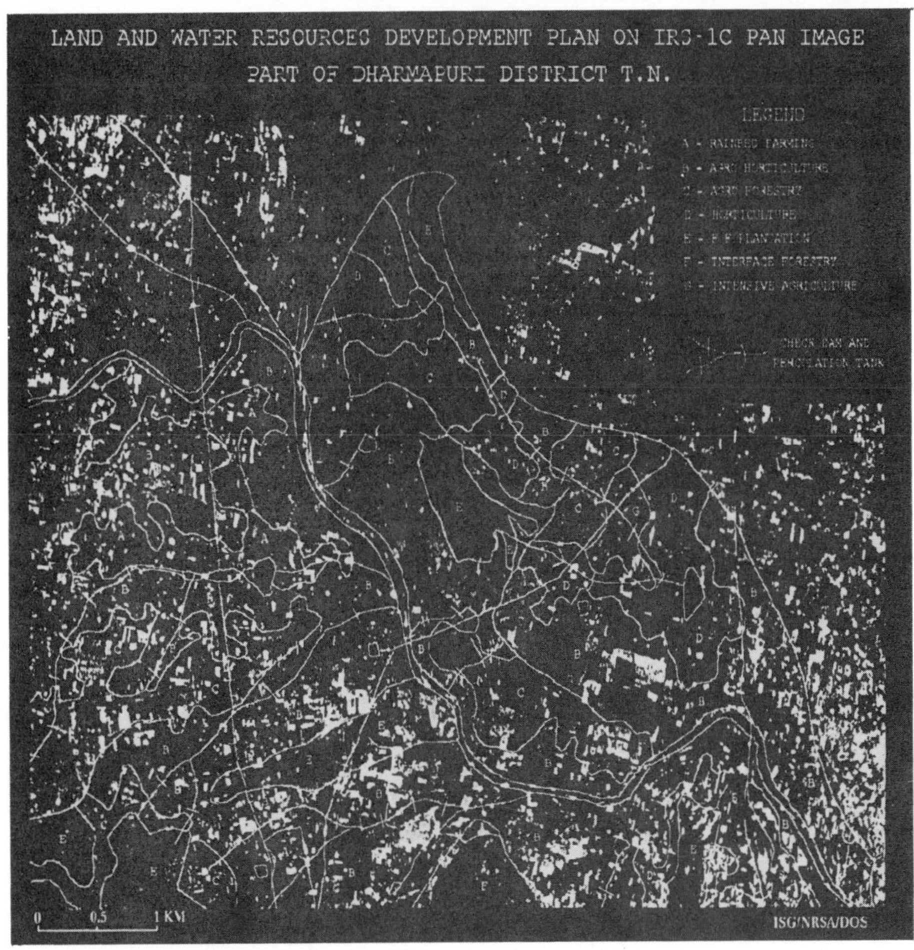

Figure 4. IRS-1C images of Dharmapuri District, Tamil Nadu

APPLICATION OF REMOTE SENSING AND GIS TECHNOLOGY IN RURAL DEVELOPMENT PLANNING IN NEPAL

*G.B. Juwa**

A. Introduction

Nepal is a mountainous country that needs information for the scientific management of land and its resources for the rural development of the country. Remote sensing and geographic information system (GIS) tools have been used successfully to study the changes in the forest resources of Nepal since the beginning of 1980. The Forest Survey Division (the then Forest Survey and Statistics Division under the Ministry of Forest and Soil Conservation) under the Forest Research and Survey Centre (FORESC) and the Forest Resources Information System (FRIS) project have been carrying out forest inventory -- both at the district level as well as at the national level -- using the latest aerial photographs and remotely sensed data together with the available topographic and land-use maps.

Information about resources for development depends on the objectives of the management system for rural development. Remote sensing and GIS have been aiding development since 1964-1965, from the aerial photographs taken in 1963-1964 for mapping and monitoring of forests. Remote sensing and GIS technology were introduced as new tools for forest inventory for the development, management and future planning in the Nepal system. Since 1990, the FRIS project of FINNIDA (Finnish government) is providing technical and financial support to the Forest Survey Division. Now, they have become very popular and handy tools for resource monitoring, evaluation and planning to meet the demands of an ever-increasing population on a sustained basis.

B. Forest resources of the country

In a hilly country like Nepal, the forest is one of the most important resources for rural development. Most of the rural people live in forests and depend on it for fuelwood, fodder, timber and income. At the same time, it is the prime ecosystem of the environment. Forests, therefore, are the most important resource, an extremely important component of the environment, and they play a vital role in the improvement of the socio-economic condition of the rural people in Nepal as well as in conserving the natural resources of the country.

The FRIS project of FINNIDA is providing technical as well as financial support to the Forest Survey Division. The objective of the project is to develop and establish a forest resource information system that can provide relevant up-to-date information about forest resources of Nepal for the purpose of development planning. Two phases of the project have been completed, and the third phase will be continuing from July 1996 for another three and a half years.

The focus of development planning in any country is to fulfil the social and human aspirations of its people, meeting the essential requirements of living, raising income levels and improving their quality of life. These days, the remote sensing section is performing activities to collect information on the different resources of the country, but mainly it is focusing on the forest sector:

(a) Prepare reports on vegetation maps of the region, using satellite images;
(b) Prepare various thematic maps of districts, using GIS technology;
(c) Classify the forests of districts, using satellite images;
(d) Prepare statistics and database maps related to forestry;
(e) Study the existing forest condition of districts, using GIS technology on satellite images.

* Forest Research and Survey Centre, Kathmandu, Nepal.

C. Methodology used in the Terai (plain) area

Satellite image analysis methods are being employed in the forest cover mapping of the Terai districts. Landsat TM (Thematic Mapper) data taken in two different years were used in the study. Images from two subsequent years had to be used because a total coverage of relatively cloudless images was not available for either year. The Land Resources Mapping Project (LRMP) maps (1:50,000 scale) based on the aerial photos of the year 1978-1979 are used for digitizing the boundary of the study area. The boundaries from the LRMP land-use maps are transferred into digital form, and then other details are added to it by using other maps, such as Indian Topographic maps of the scale 1:63,360. Normalized difference vegetation index (NDVI) thresholding is applied to separate forested land from non-forested land in the satellite images. NDVI issued elsewhere is correlated with green biomass in order to make a vital tool for the identification of forest resources. Rectification and cloud correction are done in the satellite images with the help of the LRMP land-use maps. The change in the forest areas in the plains can be obtained by comparing the forest area of the satellite images with that of the digital LRMP land-use maps. The methodology used in the Terai cannot be used in the hilly areas of the country, partly because of slopes and shades, and mainly because of the inaccuracy of the available data. However, remote sensing and GIS tools are used for the preparation of woody vegetation maps of the hilly areas of the country. The results obtained from satellite images of Terai (plain) area of the country are satisfactory, compared with results calculated using the aerial photographs of the same area.

The forest area, as indicated in the Master Plan for the Forestry Sector 1988, is 37 per cent of the total land area of the country. Unfortunately, since rural people are highly dependent on the forest resources, the resources are deteriorating. Therefore, the forests of the country should be managed in a scientific manner for their conservation and the benefit of the country's rural people. For assessing the forest resources, forest inventory in the field is the conventional way; it is a manual job, both time-consuming and tedious. The advanced technology of GIS and remote sensing is a much faster way, and high-quality work can be done.

D. Development of remote sensing and geographic information system applications in Nepal

Because it is an important tool for gathering information with which to set up a resource management system, remote sensing came to Nepal in 1981 under the name of the National Remote Sensing Centre (NRSC), which is an autonomous institution, through joint cooperation between His Majesty's Government of Nepal and USAID. It had been performing multi-disciplinary work, to generate useful information so that the technology can be applied to the national development projects. The Centre is also used as a focal point for remote sensing activities concerning natural resources. New aerial photographs of the country were taken for the preparation of the forest maps, and forest inventory in the field was carried out.

In July 1989, NRSC joined the Forest Research and Survey Centre as the Remote Sensing Section. It is one of the three sections in the Forest Survey Division under the Ministry of Forests and Soil Conservation. The achievements and activities of NRSC in the application of remote sensing and GIS are as follows:

(a) Study of deforestation for sustainable development by using satellite images;
(b) Study of the possibility of hydropower and irrigation development;
(c) Study of natural resources for management and planning;
(d) Estimation of forest areas on natural basis;
(e) Yield forecasting study.

In the last few years, the main activity of the Remote Sensing Section is to collect forest resource data using conventional and modern techniques. It is providing the information to all offices under the Ministry for sustainable management and development of Nepal's forest resources.

The various software being used at FSD to study the changes in the forest resources of the country and to prepare various maps are TOPOS, PC Arc Info and PC ERDAS.

E. Conclusion

Application of remote sensing and GIS is very useful for studying the forest resources and their changes. Maps and data are the important tools for the organization of resources where mankind's requirement and activities are continually increasing and changing. Maps with other necessary information as required by the various objectives of different projects must be updated by remote sensing and GIS technology to meet new needs. Therefore, remote sensing is playing a vital role in effective and efficient mapping and monitoring of the natural resources. Computer-based image processing systems have become less expensive and more efficient for rural development planning. The integration of a geographical information system with remotely sensed data has added a new dimension to remote sensing applications.

APPLICATION OF REMOTE SENSING AND GIS FOR RURAL DEVELOPMENT PLANNING IN BANGLADESH

*Hazifur Rahman**

A. Introduction

The ever-increasing human population over the world, with its limited natural resources, presents a continuous threat to the living standards of the people as well as to the environment in which they are living. The synergistic interaction between human activities, socio-economic conditions of the people and natural resources has a significant impact on the environment as well as on the living condition of the people. Unplanned activities and overpopulation lead to mismanagement and over-exploitation of the natural resources, which may result in irreversible damage to the environment and in a worsening of the living conditions of the populace (Verstraete and Pinty, 1991). The gravity of the problem calls for the need to harmonize the development planning utilizing the Earth's natural resources in a sustainable way to maintain a better quality of life in the present era, causing no adverse impacts for the future generations, and to maintain long-term development over a region, particularly in the rural areas because they hold a vast majority of the population in developing countries. The materialization of such an objective can only be accomplished through proper planning and management of the natural resources which requires up-to-date information on their status and availability, as well as information regarding any impending disaster.

1. Rural development planning and use of natural resources

Rural development planning is a core element in the overall development of a country, particularly in the third world regions. The threat of irreversible damage to the environment and destruction or misuse of natural resources in the rural areas can be largely minimized through proper rural development planning. Rural development planning includes coordinated action in several directions, especially agriculture extension, which includes land budgeting, crop diversification, development of irrigation potentialities, soil conservation, cooperatives, provision of inputs, infrastructure (such as the usual facilities of roads, electrification, communication, marketing, finance, storage) and others. The available natural resources play a dynamic role towards development. Such planning can ensure optimum use of all natural resources with a promise to protect the. environment and to repeatedly regenerate/restore some of the natural resources (renewable) for lasting utilization.

2. Role of remote sensing and GIS

Remote sensing, with its unique capability for synoptic viewing, real time and repetitive views, is a potential tool for rural development planning. Activities like infrastructure development, agriculture applications for crop monitoring, crop inventory, land budgeting, irrigation potentiality development, fisheries development and more can greatly benefit from remote sensing. For the management of sustainable natural resources and the environment, sequential information on their changes and development is essential. For developmental and environmental planning this information has to be assembled, manipulated, analysed and displayed in a usable form, a process that involves manipulation and analysis of huge amounts of spatial data on a regular basis. Nowadays remote sensing integrated with geographic information systems (GIS) has emerged as a highly sophisticated information technology and a useful tool to bring this objective about.

B. Rural development planning and Bangladesh

Bangladesh is a densely populated agriculture-based country with limited land area and natural resources. Moreover, the country often suffers from natural disasters caused by cyclones, droughts,

* Bangladesh Space Research and Remote Sensing Organization (SPARRSO), Dhaka.

floods and other calamities affecting the land and the people. Presently, industrialization is being attempted all over the country, although it is still constrained by large budgetary requirements, in addition to other factors. Such budgetary requirements can be largely met by the rural sector through proper planning of rural areas. Housing, communication, embankments for flood protection, shifting cultivation and so on are all leaving effects on the ecology of the country and need immediate attention. The depletion of forests in the Modhupur sal forests, Sylhet forest area, Chittagong forest area and the vast village forests is a great ecological concern. The rivers are changing their courses, developing sand bars and islands, and are drying up in places for various reasons. The groundwater table is gradually going down. Shrimp cultivation, although a profitable and growing economic activity in the coastal region of Bangladesh, has a strong impact as well. All these activities cause ecological changes in the country.

Under these circumstances, applications of remote sensing technology and GIS are increasing rapidly in this country and the Government of Bangladesh is putting emphasis on proper utilization of all sustainable natural resources, keeping a proper balance of the ecosystem. The Bangladesh Space Research and Remote Sensing Organization (SPARRSO), being the focal agency for all remote sensing and space activities, has been utilizing remote sensing and GIS for various natural resource management and environmental monitoring studies in different fields such as meteorology, agriculture, water resources, floods, forestry, fisheries, geology, oceanography, climate change and natural disasters. The outcome of these studies provides valuable information for the development planning of rural Bangladesh.

1. Application of remote sensing and GIS for development

SPARRSO has executed a number of projects on the applications of remote sensing technology in agriculture, water resources, fisheries and forestry sectors, which mostly concern the rural areas of the country. The information derived through these studies has been disseminated to the relevant implementing agencies for development planning. A brief description of some of the studies is given below.

Preparing thana *base maps:* SPARRSO prepared preliminary base maps of seventeen *thanas* (administrative units) showing features of physical resources and infrastructures. The maps were meant for assisting the *thana* engineers and technical support staff so that they would be able to use them as a planning instrument in order to plan, monitor and maintain rural infrastructure development effectively for annual development projects.

Monitoring forest cover: Under this project, remote sensing techniques and GIS have been used to map and monitor the forest covers of the Chittagong Forest Division, using multitemporal Landsat-MSS and TM data. This study provides valuable information on the forest resources and their changes over the study areas.

Mangrove afforestation project: SPARRSO was assigned the task of monitoring, by remote sensing, the mangrove afforestation in the costal area of Bangladesh carried out by the Forest Department between 1985 and 1990. A total of 192 maps (scale 1:10,000) were prepared showing old plantation areas with their present status and newly accreted/eroded land, along with identification of areas for new plantation. A total of 45,087 acres of mangrove were planted successfully by the Forest Department during the second phase to improve the ecology of the coastal region. This second phase was a follow-up of the first phase (1980-1985), in which 88,203 acres of plantations were raised in the coastal belt of Bangladesh.

Monitoring changes in river courses: The random shifting of the river courses (namely the Padma, the Jamuna and the Tista), due to the continuous process of erosion and sedimentation, renders people homeless and destitute. Navigational, irrigational and flood problems also crop up. Under this project, multitemporal data acquired from satellite platforms for the period of 1973 to 1990 were used to analyse and study the changing behaviour of the river courses in relation to the erosional/acrretional activities between consecutive years.

Upgrading inventory of inland water bodies: Under this study, different types of water bodies in the country were identified and classified using infra-red aerial photographs, SPOT and Landsat TM. Potential areas of fish culture and ice plants for preserving the fish were identified with Arc Info GIS and, in addition, chlorophyll content was estimated. This study included mapping of the small water bodies (ponds and lakes of less than 25 hectares) and determining their status in 40 *thanas*.

Application of GIS for fisheries: A study was carried out using remote sensing and GIS facilities to select suitable sites for coastal shrimp farm development in the south-western part of Bangladesh. The digitized layers created will be analysed using up-to-date TM imagery; some field work is required for the completion of the study.

Remote sensing study on irrigated area assessment: A research project was undertaken by SPARRSO to assess the irrigated area over the north-eastern and central part of the country using remote sensing and GIS. The analysis showed that there was a distinct rise of boro cultivation over the majority of the study area during the last seven years. The study will help the planners to assess the present and past status of irrigation for winter rice cultivation and to plan the future activities in these regards.

Crop growth monitoring: Under this research project, a quantitative approach has been taken for monitoring crop growth through estimation of different biophysical parameters characterizing the two major crops in Bangladesh, namely wheat and rice. The approach consists of retrieving the biophysical parameters of the crops through inversion of a surface bidirectional reflectance model against directional measurements over the two canopies. The preliminary results of this study are encouraging and the technique provides a reasonably accurate estimation of the biophysical parameters.

Flood monitoring and flood plain mapping: In addition to early warning, digital techniques for flood monitoring and flood area mapping using satellite data are being developed at SPARRSO. The floods of 1987 and 1988 were monitored at SPARRSO using real-time NOAA-AVHRR data, and the processed images of floods were distributed to various government departments and other organizations for taking precautionary measures.

A pilot study has been made in collaboration with the Chinese Academy of Sciences on a technical cooperation among developing countries (TCDC) basis with support from UN/ESCAP for flood plain mapping and flood monitoring for a selected site using remote sensing technique and GIS.

Remote sensing monitoring of water quality of large water bodies: This research work has been undertaken for monitoring water quality of large water bodies in Bangladesh using remote sensing data through the development of a water quality model. The principal parameters of interest are sediment content, salinity and chlorophyll concentration.

Aerial photography project: Under this project, aerial photographs of the Sundarbans, Chittagong and Cox's Bazaar forest divisions have been acquired for the purpose of forest inventory and mapping, which are ultimately providing a reliable data base for forest resource management and planning.

National Conservation Strategy (NCS): The need for a strategic approach to natural resource use through sustainable development has been realized; consequently, the Government of Bangladesh established the National Conservation Strategy (NCS) secretariat at the Bangladesh Agricultural Research Council (BARC) to assist the government in the preparation of NCS. Under the NCS programme, SPARRSO and Survey of Bangladesh have been assigned to conduct the survey for the preparation of base maps and land-use maps of the study area.

C. Conclusion

Rural development planning is an essential step towards the overall development of a country, a step that, it is hoped, will ensure agreeable living conditions for the people without upsetting the environmental or ecological balance. Proper development planning requires thorough knowledge about the available natural resources and their status, as well as a knowledge of the physiographic environment so that infrastructure systems may be properly developed. In this regard, integrated remote sensing and GIS play a vital role. However, the potentialities of these tools are still to be exploited and it requires further investigation to take advantages of the recently developed technology.

The exchange of ideas between different countries regarding the application of remote sensing and GIS can greatly enhance the region's ability to manage the ecosystem, and thereby development can benefit and be accelerated.

However, in order to take full advantage of the recently developed technologies, good coordination between the planners/decision makers and remote sensing technologists is highly desirable. In this regard, professional planners at all levels should be made aware of the potential usefulness and capabilities of remote sensing and GIS in planning and development activities of a nation.

Bibliography

Annual Report of SPARRSO, July 1993-June 1994.

Annual Report of SPARRSO, July 1992-June 1993.

Verstraete, M.M., and B.P. Pinty (1991). The potential contribution of satellite remote sensing to the understanding of arid lands processes. *Vegetatio*, 91: 59-72.

POLICIES AND PROGRAMMES FOR RURAL DEVELOPMENT IN INDONESIA

*Kawit Widodo**

A. Introduction

Development in the First Long Term Development (FLTD) has succeeded in increasing the living standard of people including that of rural communities. Nevertheless, rural development in the Sixth Five-year Development Plan (SFYPD) and the Second Long-term Development (SLTD) should be continued and accelerated as the majority of the rural population is still left behind compared to the communities in the urban areas. The main challenges faced in rural community development in the SLTD, particularly in the SFYDP are to alleviate rural proverty, improve the quality of human resources, optimally utilize natural resources for increasing food production and industrial raw materials, and develop an interrelatedness between rural and urban areas for mutual support and benefit.

In response to the challenges, there are some constraints, among other things the lack of capability of village administration in providing proper village infrastructure and facilities, lack of human resources in the rural areas for implementing development, and lack of capability of rural institutions.

Besides the aforementioned constraints, there are some opportunities capable of being developed in rural development, such as the achievements of development in the FLTD, unutilized potential natural resources, strong rural community self-help, and information quickly reaching the rural community.

B. Targets and policies

1. Targets

(a) Targets in the SLTD

The main targets of rural development in the SLTD are to create a strong economic condition of the rural community that may grow and develop sustainably, achieve interrelatedness of urban and rural economy, create prosperous rural communities, and alleviate rural poverty.

(b) Targets in the SFYDP

The targets of rural development in the SLTD will make efforts to achieve goals in phases, starting from the SFYDP. The development will be accelerated in the SFYDP, characterized by improving the quality of human resources in the rural areas, viewed from the levels of community well-being, education and skills capable of motivating initiativies and self-help of rural community; strong economic structures characterized by improving diversified business activities producing various local prime commodities as well as supporting proper rural economic facilities and infrastructure, improved awareness of rural community regarding environmental-concept development; better functioning of the community-based institutions and village administration in improving effective implementation of rural community development; and decrease of the number of the rural poor and less-developed villages.

2. Policies

Within the framework of achieving the targets mentioned above, rural development policies in the SFYDP are formulated by improvement of the quality of rural employment, improvement of

* Directorate-General of Rural Community Development, Jakarta, Indonesia.

capability of the community in production, development of rural infrastructure and facilities, institutionalization of integrated area development approaches, and strengthening of village administration and village-based institutions.

(a) Improvement of the quality of rural employment

Policies on improvement of education and skills for rural employment are directed to improve rural human resources in order to be able to take active participation in development, be able to take opportunities in large-scale business activities, be able to increase production, processing and marketing, and be ready to enter the work market, especially in the urban areas, leading to an increase in the welfare of rural community.

Policy measures on improvement of the quality of rural employment are improvement of primary and secondary education, as well as vocational facilities, in accordance with the potentials and the needs of the community; expansion of education and vocational training for diversification and processing of production in accordance with market standards and demand; improvement of capability of organizing rural communities in a vehicle that is cooperative in nature; implementation of effective guidance and counselling through rural community business groups; and improvement of health services for rural communities.

(b) Improvement of community capability in production

Policies on improvement of community capability in production are directed to materialize a stronger rural economic structure, based on a state agricultural sector that can maintain self-sufficiency in food, diversify and supply food stuffs, and support the development of rural industries.

Policy measures to be taken to improve capabillty in production are development and application of appropriate technologies in improving capacity, quality, and adding value in production; improvement of capital facilities and access of comnunity to capital by promoting cooperative, economic, and other finance institutions; and improvement of agricultural production, processing and marketing facilities.

(c) Development of rural infrastructure and facilities

This policy is directed to create functional interrelatedness of inter-village groups and of villages and towns, eliminate the isolation of villages, and accelerate the development of social activities and rural economy, as well as satisfy the needs of rural community for basic facilities and infrastructure.

Policy measures to be taken, among others, are improvement of rural transportation facilities; expansion of rural electrification services; development of production centres and marketing of agro-industry, wood industry and other community industries; improvement and equitable distribution of health, education and environmental sanitation facilities; and development of rural irrigation networks.

(d) Institutionalization of integrated area development

This policy is directed to implement rural development in accordance with problems faced by rural communities through the integrated area development approach. Development of a rural area, through this approach, will be handled sectorally. Implementation with active participation of the local community will start from planning to implementation and monitoring. Besides the sectoral approach, implementation of integrated development is carried out in the villages as inter-village development centres. Within the framework of its effective implementation, there is a need to identify villages having potential for intensifying area development.

Policy measures to be taken, among others, are implementation of integrated development, particularly in poverty alleviation; and development, as well as institutionalization and coordination function in planning, implementation and monitoring of rural development at village and sub-district levels.

(e) Strengthening village administration and village-based institutions

This policy is directed towards effective and efficient implementation of rural development so that the authority and responsibility of village administration in managing the development may be more intensified. This policy is also aimed at improving the role of village-based institutions such as the Village Community Development Council (LKMD), Village Development Cadres (KPD), the Family Welfare Movement (PKK), Natural Conservation Cadres, and Natural Resources Preservation Groups (KPSA), so that the community may take more active participation in development.

Policy measures to be taken are improvement of capability of village administration apparatuses in development and administration, improvement of village administration structures in accordance with concerned village development, improvement of capability of village administration in discovering village fund resources, provision of village administration facilities and infrastructure, development of information systems supporting village development planning, and improvement as well as strengthening of capability and function of village-based institutions in accommodating and channeling the aspirations of rural communities.

C. Development programmes

There are seven rural development programmes in the SFYDP.

1. Improvement of community education and skills

Programmes on improvement of community education and skills cover (a) implementation of compulsory education of nine years and eradication of illiteracy through implementing learning groups of package A and B, (b) improvement of community skills, (c) provision of extension/field workers either from government or non-government officials in the areas of production, processing and marketing of goods and services such as agriculture, forestry, mining, small industry, trading, tourism and the like, (d) development of skills and education programnes for development of market-oriented local economic activities and (e) extension for rural communities in the framework of preservation of rural environment. Priority of the programme will be the children of school age and youths, as well as the drop-outs, particularly those who live in less-developed villages.

2. Improvement of community health

Programmes on improvement of community health consist of (a) improvement of community nutrition through diversified food and extension on healthy living, (b) intensification of environmental sanitation and (c) intensified activities of integrated health post. Priority of the programme will be the health of pregnant women and under-five-year-old children, especially those who live in less-developed villages.

3. Development of appropriate technology

Programmes on development of appropriate technology cover (a) development and application of technologies capable of accelerating the growth of rural agro-industries, (b) development ond modernization of integrated farming activities and (c) development and utilization of environment technologies of clean water supply, sanitation and settlement, among others.

4. Mobilization of community participation

The objective of the programme is to mobilize participation of youths and women in development activities through extension and counselling, and implementation of development in rural areas, as well as improvement of skills of rural low-income communities. This programme is also carried out through activities of village-based institutions such as Village Community Development Council (LKMD), Family Welfare Movement (PKK), youth organizations, Scouts and cooperatives.

5. Development of rural facilities and infrastructure

Programmes for development of rural facilities and infrastructures consist of (a) development of communication facilitles and infrastructure covering roads and bridges, transportation facilities, quays, as well as development of rural electrification and postal services, (b) improvement of rural clean water supply and (c) improvement of health and education facllities and infrastructure.

6. Strengthening rural institutions

Programmes for strengthening rural institutions cover (a) improvement of capability of village administration apparatuses in managing development, followed by rehabilitation of village administratiom facilities and infrastructure, (b) strengthening of the function and the role of LKMD, (c) improvement of PKK's programmes as a vechicle for women's activities in development, (d) strengthening of the function and the role of KPD, Natural Conservation Cadres and KPSA in guiding and organizing the community, (e) promotion of Scout and youth organizations in preparing the youth to participate in development, (f) strengthening the mechanism of Local Development Working Units (UDKP) as village development coordination at the sub-district level and (g) development of other social services for rural communities.

Some of the ways that rural economic institutions are strengthened are (a) increased capacity and services of Village Unit Coorperatives (KUD) in providing credit and production marketing, (b) guidance and training for improving capabilities of community groups, (c) improvement of the mechanism of credit supply in order to promote accessibility to financial resources, (d) special assistance for development of community economic activities, (e) improved procedures and mechanism of the role of the private sector in development of the rural economy and (f) improvement of land reform and certification of agricultural land.

Besides the programmes mentioned above, the following progrommes will also be implemented according to the village improvement stages. Programmes to be implemented for fast-developing villages are directed particularly towards (a) improvement of services of rural financial institutions, for small-scale industries in particular, (b) education and skill programmes for development of market-oriented local economic activities and (c) provision of market information and other general information for communities to enter fields of work in non-agricultural sectors.

Less-developed villages will receive special attention through such programmes as (a) development of human resources, (b) improvement of community health, (c) application of appropriate technologies in community activties, (d) strengthening such rural institutions as KIUD, Village Credits and the function of government, as well as community-based institutions, (e) development of basic and economic facilities, particularly transportation facilities and (f) special assistance for development of community economies.

7. Development of village mapping (*sketsa*)

Programmes to develop village mapping are conected with village planning development. They consist of (a) the borders between villages, (b) legend of mapping (*sketsa*) and (c) space planning.

Geographic information systems (GIS) are very important for developing village *sketsa*, especially for village planning.

USING INTEGRATED REMOTE SENSING AND GIS IN SUSTAINABLE RURAL DEVELOPMENT

*Tran Thi Kieu Hanh**

A. Introduction

Viet Nam is an agricultural country where about 90 per cent of the population are living as farmers. Therefore, economic agricultural development is the first goal in the socio-economic development strategy of Viet Nam's government. In the rural areas of Viet Nam, the paddy fields are distributed separately and wetland rice is traditionally cultivated. Nowadays, because of the population explosion, there are some problems such as food and water shortage, deforestation, erosion, environmental pollution, social evils and so on. Therefore, sustainable rural development has to be paid serious attention in Viet Nam.

Sustainable development means reasonable utilization of natural resources and environmental protection. Both aspects -- development and protection -- have to combine regularly and interactively. In order to solve the above problems effectively, it is necessary to use integrated remote sensing and geographic information systems (GIS) to aid sustainable rural development.

The integration of remote sensing and GIS is a considerable reality for natural resource management in the rural areas. The main potential benefits of integrated remote sensing and GIS are in updating map information, improving thematic mapping using remotely sensed data, analysing data, and modelling, in order to create the final products for exploitation of natural resources and for environmental protection. In the case of sustainable rural development, some of the territorial planning projects for reasonable utilization and environment of natural resources of the rural areas employ the new technology to come up with a wide variety of maps. Concretely, depending on the objectives of the projects, there can be some resource-assessment maps on rural socio-economic development accompanying the tabular information, such as soil evaluation for industrial crop or food crop, environmental monitoring of land use, forest, erosion or flood, or other types of resource-assessment maps.

The integration of remote sensing and GIS for sustainable rural development has some benefits:

(a) Cost-effective, routine updating of natural-resource map information, because remote sensing is often the most cost-effective updating procedure (changes in land use, forest, vegetation cover, erosion types and so on);

(b) Improved accuracy of computer-assisted techniques for information extraction from digital images (incorporating ancillary data such as slope, hydrology and climate parameters);

(c) Inclusion of digital images like ortho-photographic maps into GIS as a data layer;

(d) Prediction of future scenarios (especially the impact of future changes in crop production);

(e) Geometric correction of remote sensing digital imagery.

B. An example of using integrated remote sensing and GIS to make the plans for socio-economic development of Co To District (Quang Ninh Province)

1. Study area

Co To District was set up only recently. Formerly it was part of Cam Pha District (now Van Don District). Because of many changes in the political, social and economic situation, the natural

* Institute of Geography, Viet Nam National Centre for Science and Technology, Hanoi.

resources of Co To are greatly reduced in both quality and quantity. Before 1980, Co To was rich in forests, rice and fish. Co To was well known for its fish production. But now the forest area is reduced to only 25 per cent of its previous size. Crop land is also reduced, as well as crop production. There are many environmental problems occurring: lack of water for agricultural and domestic use, soil erosion, soil degradation and sand-bearing wind, to name only a few. So it is necessary to make a plan for socio-economic development of Co To.

2. Methodology

There are some methods for resource assessment, but the best is integration of remote sensing and GIS, since it is cost-effective (saving both money and time) and accurate. The need is especially urgent in Co To District, where natural resources are changing from day to day.

Remotely sensed data was used as input for GIS, including aerial photographs from 1960 to 1995, and Landsat TM imagery from 1992 at a scale of 1:50,000. The software used is MapInfo version 3.0.

Visual interpretation of remote sensing data for some thematic maps was compiled: land use, vegetation cover, soil type, hydrology, hydrogeology and geomorphology. The thematic maps were used for the spatial data in the database. Resource inventories were used for compiling the attribute database in the form of tabular information: area, slope, elevation, population, crop yield and water quality and quantity.

3. Results

During analysis of the data, different maps were overlaid to create many new project objectives and the database developed. For instance, geomorphological and soil-type maps were overlaid to look for land suitable for crop cultivation or some type of industrial crop land; also, geomorphological, slope, and vegetation-cover maps were overlaid for an afforestation project and protection of forest and soil erosion potential. In this case, the complex map to be used in the territorial planning project was the final product.

C. Conclusion

Using integrated remote sensing and GIS for studies of sustainable rural development in Co To District is necessary and useful, especially in cases of the investigation, assessment and management of natural resources. This method greatly helps the local government and planners in planning, zoning, property assessment and the rational use of the rural resources and environment. Rural resource managers rely on this method for fish, wildlife planning, management of forest, agricutural and other resource management tasks. This technology is very cost-effective when it comes to routine updating of map information. But it demands an investment in hardware and software. However, it also depends much on skilled people who can operate computers, who are conversant with the software, understand the data types that are stored and can relate to the problems of the environment and natural resource management.

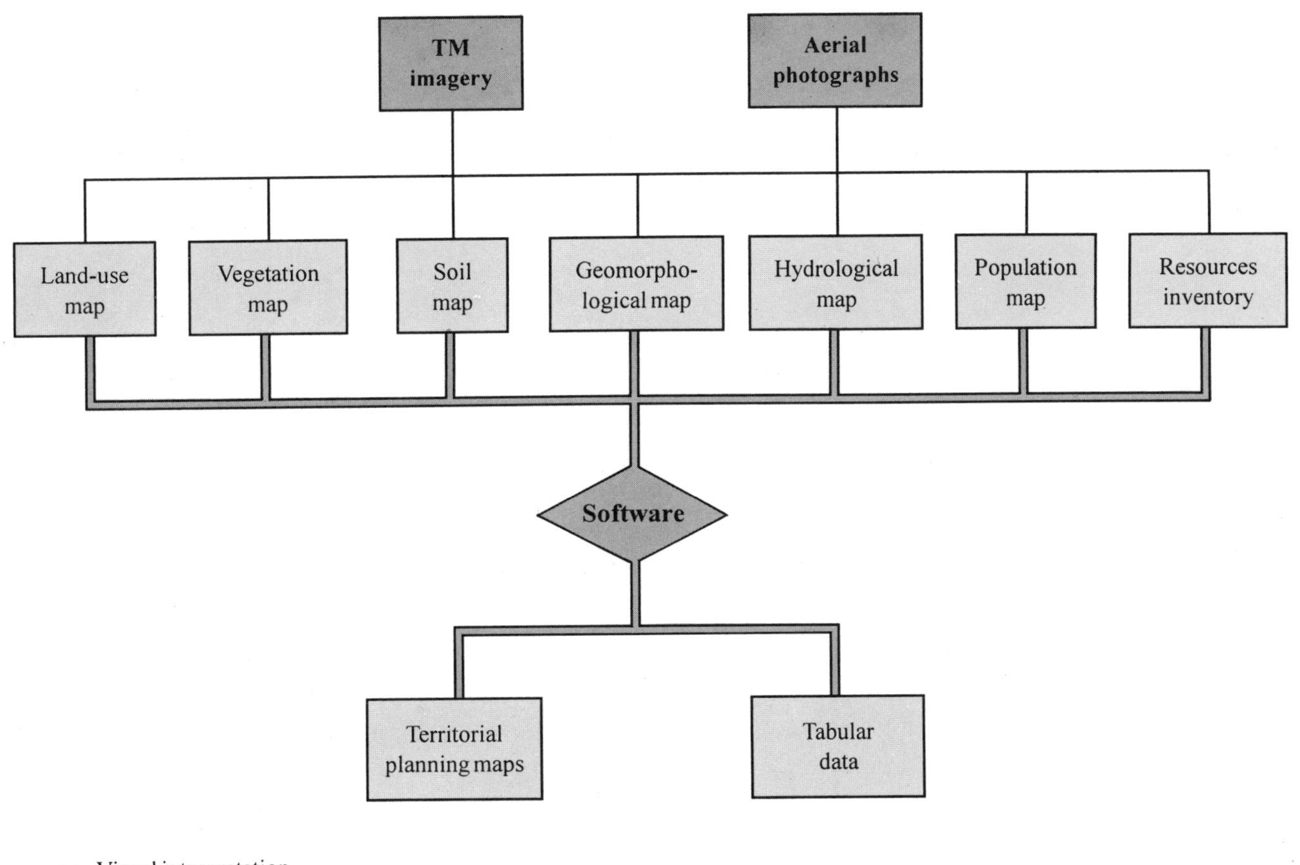

——— Visual interpretation.
═══ GIS operation.

Figure 1. Block diagram of the approach taken to make some territorial planning projects of Co To District using integrated remote sensing and GIS

REMOTE SENSING APPLICATIONS FOR ENVIRONMENTAL MANAGEMENT WITH SPECIFIC RELEVANCE TO RURAL DEVELOPMENT

*C.B.S. Dutt**

A. Introduction

The remote sensing of the environment is of great importance because of the concern for environmental management problems in India, which arises from a number of causes, viz., increasing human and animal population exerting pressure on natural resources, general indifference in the industrial sectors to environmental safety, and their phenomenal growth leading to poor ambient air quality and indiscriminate land management practices. This has resulted in economic growth corresponding to requirements by expansion of availability of resources and energy. The energy mix in the country is such that India has been rated as low in environmental literacy, thus creating a gross undervaluation of the economic and ecological aspects in various sectors.

Since independence India has achieved remarkable progress in raising the agricultural production from 50 million tons to 190 million tons in 1995. Similarly, for example, sugarcane production has increased from 5,100,000 tons (1980-1981) to 11 million tons in 1990 (Swaminathan, 1991). Having made such remarkable achievements in boosting agricultural production, we are left with ecological implications that have been accumulating all through these years, causing concern for conservation strategists.

The prevalence of drought in most parts of India may be attributed partly due to the failure of the monsoon. However, if we look at India's water potential, India receives a fairly good amount of rainfall (average annual rainfall is 100 mm) and has vast water potential. The net sown area in the country is about 141 million hectares. The weighted average requirement of water per ha works out to be 700 mm (depth), and on this basis it is possible to provide irrigation facilities to 153 million ha, i.e. the entire cultivated area of the country (141 million ha) for at least one crop. Yet in most of the areas the crop productivity "per unit water" is the lowest in respect of paddy and sugarcane compared to irrigated dry crops (*The Hindu,* 1991; INAE, 1990).

Of late, India has been facing a transition from production-oriented schemes to ecologically sustainable schemes in every resource sector. The cumulative impacts of the series of environmental perturbations unleashed by the overgrowth of population and the associated demands of consumerism on the biosphere are presenting unbridgeable gaps in the supply of resources.

The present per capita amount of land in India is about 0.17 ha and by 2010 AD it may decline to 0.15 ha. By the end of 2010 AD we may require more area under agriculture, from the present 141 million ha to 160 million ha, in order to meet the food grain requirement with a target of 225 million tons, at an annual growth rate of 6 per cent. However, the present growth rate is estimated at 3 per cent per year. If we optimistically enhance the agricultural production on an ecologically sustained basis, we will be able to achieve this by doubling the growth rate production annually.

B. Biological diversity

With the consequences on environmental problems that mankind will face in the future, it might lead to the loss of at least one higher plant species per day from tropical forests alone. Estimates indicate that from 10 hot spot localities in tropical forests alone, 17,000 endemic plant species and

* Forestry and Ecology Division, National Remote Sensing Agency, Hyderabad, India.

350,000 endemic animal species could well be eliminated shortly. If the present trend continues, about 25 per cent of the total 250,000 higher plant species, which constitute roughly 16 per cent of the total global plant wealth, could be lost, and it is feared that 15-20 per cent (i.e. over 2,500 species) of the total vascular flora now fall in the category of threatened species. *Therefore, India has been considered as the most diverse of biological communities and recognized as one the world's top 12 mega-diversity nations* (Madhav, 1991).

There are no clear estimates about the marine biota along the Indian coastline, which is 7,500 km long with a shelf zone of 452,460 sq km and extended economic zone of 2,013,410 sq km (Khoshoo, 1991). There is an abundance of sea weeds, fish, coral and other marine life. From the biosphere point of view the marine biota and the associated environment are little understood and need attention.

Although biological extinction has been a natural phenomenon in the past, perhaps at the rate of one species for every 1,000 years, the rate of extinction has never been as rapid as it is today, because of environmental degradation. Hence this is an area of concern in spite of conflicting statistics. The use of remote sensing data could be employed for large areas to delineate zones based on fragmentation, in order to "prospect" for biodiversity zones.

C. Biosphere reserves

In India there are 54 national parks and 372 sanctuaries with a total area of 109,652 sq km. This number will increase with future additions. These include 13 biosphere reserves for which detailed information about their habitats and the holdings have been documented. India has done incredibly well in identifying the zones for environmental protection based on biospheres (Government of India, 1987; Khoshoo, 1991). However, there is a necessity to document and inventory the resources with respect to flora and fauna periodically to account for the changes that are taking place. Obviously, the study of such large biogeographical zones is made possible only by use of remote sensing.

D. Issues of environment

Several developmental projects are taken up in several parts of the country either for industrial or for power sectors. Whatever the developments may be, the impacts are leading to deterioration of environment in the area or surroundings. Because of the lack of an adequate database on the pre-establishment stage, developmental stage and the post-development stage, environmental impact studies with respect to every sector need to be studied in an integrated manner attaching top priority to environmental conservation. With the increasing damage to agriculture, quality of life in terms of ambient air quality clearance of forest due to hydropower projects, increased urbanization, industriali- ation, mining and other activities, the country needs an alternative strategy suitably to compensate the losses without causing much difference to the already existing environment and the available natural resources.

E. Ecological "hot spots"

Thirteen biosphere reserves have been selected in India; they are considered as virgin areas for preserving the already existing flora and fauna without making further drains on the germ plasm. However, the conservation strategy needs also to be concentrated on the "hot spots", where the ecological degradation is rapidly occurring. It is strongly recommended to identify the hot spots in all the agroclimatic zones of India and declare them to be ecologically sensitive zones; moreover, they should be taken up on top priority basis for conservation/reclamation strategies. Remote sensing can play a vital role in keeping these areas ecologically sustainable.

F. Wetland environment

India has a vast extent of wetlands, wherein the mangroves alone cover an area of 6,740 sq km, which constitutes about 7 per cent of the world's mangroves. According to a recent estimate of the Department of Environment, there are 2,167 wetland areas available in the country. The unique characteristic of wetland is the presence of water and a water-saturated soil. This may be either a persistent feature or occurring for a part of the year. This ecosystem is a zone between the terrestrial and the aquatic systems, characterizing a unique environment. Diversity of character, size, type and mode of occurrence of wetlands have given rise to diversified biota. These ecosystems are useful for nutrient recovery and recycling, releasing excess nitrogen and toxins, and absorption of heavy metal residues through the tail ends of waste water channels. The fragile and unique wetland environment needs to be intensively studied and the area-wise conservation plans to be chalked out to preserve these little-attended but most important, fragile systems (Mong, 1991). Satellite remote sensing data has been used to map the coastal wetlands and a few major wetlands in the country for conservation plans.

G. Degradation of ecosystems

The alarming rate of land degradation has immensely altered and degraded the ecosystems. Because of soil erosion, water pollution, air pollution and deforestation, it is estimated that in India 5,300 tons of soil is eroded every year, that is, about 16 tons per hectare, along with major nutrients, nitrogen, phosphorous and potash. The equivalent soil loss is estimated to contribute to the loss of 30 to 50 tons of annual crop production (Venkataramani, 1991).

It is estimated that 3 to 15 million ha of tropical forests are cleared every year (about 1 acre per second). Overgrazing, deforestation and faulty land management practices have led to unprecedented devastation, causing extensive land degradation like gullying, land slips and flash floods. For example, an average of 25 million ha forest area could hold water equivalent to that of 2.5 times the storage capacity of Bhakra reservoir (7,191 m cu m). India is losing forests at the rate of 0.5 million ha per year, which means flash floods and the associated soil erosion. Shifting cultivation in India is practised in 4.23 million ha in the country; so far about 127 million ha have already been affected in north-eastern India, out of which only 32 million ha have been treated. Remote sensing data in India is being used to monitor the shifting cultivation and this will help in providing compensatory rehabilitation areas for the local people or to evolve new sustainable land management practices to prevent damage to forests as well as to reduce soil erosion. The application potentials of remote sensing in forestry environment and management with emphasis on rural development are discussed hereunder.

H. Environmental applications for sustainable development

1. Forest cover mapping and surveillance

The increasing population is exerting pressure on forests for fuel, fodder and timber. Consequently India's forest is undergoing rapid changes leading to forest degradation. The problem is further accentuated due to increased activities of mining, industrialization and construction of multi-purpose river alley projects. This has necessitated nationwide monitoring of forest cover to make strategic planning and to maintain adequate forest cover as prescribed by the Forest Act in 1956. Though the Forest Act has prescribed 33 per cent of the forest cover to the geographical area, now India has only 22.7 per cent forest cover on official records under the jurisdiction of the Forest Department.

Towards the goal of nationwide monitoring, remote sensing technology is being used to map the entire country's forest cover for the periods of 1972-1975 and 1980-1983, updated to 1981-1983. The methodology has been jointly evolved by the National Remote Sensing Agency (NRSA) in the Department of Space, and the Forest Survey of India (FSI) in the Department of Environment. The status of forest cover in India is shown in table 1.

Table 1. Status of forest cover in India

(in million hectares)

Forest type	1972-1975 %	1981-1983 %	1985-1987 %	1987-1989 %	1989-1991 %
Dense forest	46.55 (14.12)	36.14 (10.99)	37.84 (11.51)	38.50 (11.71)	38.55 (11.72)
Open forest	24.48 (7.38)	27.65 (8.41)	25.74 (7.83)	24.99 (7.60)	25.04 (7.61)
Mangrove	0.10 (0.30)	0.40 (0.12)	0.42 (0.13)	0.42 (0.13)	0.42 (0.13)
Total	71.03 (21.60)	64.20 (19.52)	64.01 (19.47)	63.92 (19.44)	64.01 (19.47)

The method of national level forest cover mapping has been operationalized, and biennially the national level forest cover mapping is being done by FSI using IRS data.

2. Grazing resource assessment

Grazing lands are important ecosystems, which support 200 million people all over the world for some form of pastoral production, of which nearly 40 million people are dependent only on livestock. In India 12.5 million hectares of land (3.7 per cent) has been recorded as permanent pasture and grazing land. However, accelerated growth in livestock population and degradation of pastures led to shrinkage of grazing resources. This has resulted in reduction of the land's carrying capacity, to the detriment of animal husbandry, which contributes 5-6 per cent of India's national income. In view of this problem, considerable attention needs to be paid to the grazing resource's potential and the need for better fodder management plans.

Satellite remote sensing data analysis procedures have identified open grasslands, shola grasslands, banni grasslands and alpine pastures. In addition, the savannah grasslands in the protected areas have provided better understanding in utilizing remote sensing technology to extend the productive potential of existing grazing resources. However, the inclusion of pasture lands and the annual productivity of pastures for organized area requires plans for the grazing resources.

India can produce 16.8 tons of green fodder per annum per hectare. This amounts to 588 million tons per year. Assuming that all 35 million hectares of wasteland is used properly as pasture land, even at this rate it can sustain only 330 million domestic animals, whereas the country has 400 million domestic animals. This large gap can be filled only by organized fodder resource planning, integrated with wood forage biomass production and mixed tree systems. Remote sensing data utilization can provide information required for achieving better solutions for sustainable management of grazing resources.

3. Fuelwood resource assessment

Out of the 900 million people in India, 70 per cent of them live in the rural areas. Most of the population depends largely on firewood for their day to day energy requirement. It has been estimated that the country's annual firewood requirement is about 150 million tons. The present capacity of firewood production, from the forests as well as non-forested areas, the official estimates record at 70 million tons. There is, then, a gap of 80 million tons of firewood requirement, which is the main cause of excessive deforestation and degradation of forests. India has 35.6 million ha of wastelands potentially available; assuming that firewood-producing plantations are increased by 50 per cent, with these areas then it would be possible to meet the demand -- provided 5 tons per hectare productivity is targeted. This is a formidable task, to bridge the gap of energy disparity through organized rural energy plantation programmes, but it would solve the ever-increasing basic requirement for firewood in rural areas.

A model study using remote sensing data on fuel wood sustainability in rural areas around south Indian forests (Narasapur, Andhra Pradesh) revealed that 68 per cent of the annual increment in forests is used as firewood, whereas 4.7 per cent comes from agricultural residues, and 3.7 per cent from animal waste. The study extended the scope and approach on fuel wood sustainability assessment by the use of remote sensing and ground data.

4. Forest non-timber resources assessment

India's forest cover constitutes 19.47 per cent of the total geographical area (1993) which is about 6,400,000 sq km. Of this, 11.73 per cent is the dense forest and 7.61 per cent is the open and degraded forest. It is estimated that per capita recorded forests of India is only 0.09 hectares when compared to world average of 0.89 hectares. The mean growing stock of the country is 74 cu m per hectare, which represents poor productive potential as a whole. The annual increment is 52 cu m, which is about 1.2 per cent of the total growing stock. The average annual production of wood per hectare is 0.07 cu m, which compares poorly with the world average of 2.1 cu m. The natural regeneration in Indian forests is also virtually negligible in about 52 per cent of the forests.

In the light of poor productivity and lack of regeneration capacity, it is necessary to harness maximum returns, in terms of economic contribution and resource requirement, for the local people. While there is 74 cu m per hectare growing stock available, the extractable non-timber resource in the form of branch wood and deadwood foliage biomass would contribute to nearly 20-30 per cent of growing stock. This would even be more in the case of moist deciduous and dry deciduous areas. The planning of forest non-timber product assessment and organized planning would reduce the demand on firewood requirement immediately for the adjoining inhabitants. Satellite remote sensing -- while it provides growing stock measurements through various sampling procedures and the development of non-timber functions for each of the locations -- would enhance forest management and generate basic firewood resources for rural development in sustainably managed forests.

5. Non-forest timber resources assessment

India has 141 million hectares of arable land and about 35 million hectares of wastelands. These areas are constituted by various degrees of tree cover, either in the form of woodlots, avenue plantations, orchards, tree plantations or scattered trees. Most of the productivity of these unaccounted non-forest timber resources (NFTR) is exploited indiscriminately. The extent at which these resources are depleting and their productivity is rather unknown. The use of high resolution images, like IRS-1C PAN data with its 5.8 m resolution, facilitates in stratifying the area and thereby, with suitable ground sampling, it is possible to estimate the non-forest timber resources. The systematic area planning of NFTR provides better insights into the locale-specific supply potential and enables the Government to make plans to bridge the supply-demand gap for the local inhabitants in terms of marginal wood or firewood requirement.

6. Afforestation programmes

India has 35.64 million hectares of underutilized lands/wastelands that have been mapped using satellite data in the 241 critically affected districts. In order to enhance the fuel and fodder requirements, if 50 per cent of the available wastelands were converted for the fuel wood crops or for pasture development programmes it would contribute to the rural economy and to the overall rural development. The use of satellite technology has already proven its value in successfully identifying the wastelands, and with the availability of high-resolution IRS-1C satellite data it is possible to monitor the afforestation programmes and to suggest areas needed for conservation or protection. The area-wise monitoring of such programmes is primarily intended for rural resource accounting, which is feasible through remote sensing and GIS procedures for successfully monitoring the plantation programmes and taking decisions on the development.

7. People's participation and JFPM activity

With the increased pressure on forests all over the world as well as in India, forest management has shifted its priority from production forestry to conservation forestry. In light of this change, the scientific management of forests and categorization of areas into different zones have been adopted by the Karnataka State forest department as part of Western Ghats eco-restoration project. Towards this, NRSA and Karnataka Forest Department jointly carried out a project in an area of 10,280 sq km in Uttar Kanara forest circle by generating multi-thematic information at 1:25,000 scale using aerospace techniques. The thematic information generated includes forest density and height maps, forest type maps, slope and aspect maps, drainage maps up to first-order channels, and volume class maps, besides consultation of contour maps with the available topographical sheets. The availability of such an enormous database, when fed into the geographic information system, enhances interpretability of the data and enables foresters to classify the area into different zones required for effective forest management practices on a scientific footing. Accordingly, the forests are proposed for categorization of zones for specific management practices. Zoning is the process of describing the physical consequences of the management plans that have been understood for sustainable forest resources in the area. It is essentially a link between management objectives and physical operations on the ground. Hence, using multi-thematic information, the area studied has been recommended for zoning, as shown below in table 2, employing the knowledge base for effective forest zoning in the model area.

Table 2. Forest zonation matrix

	D	H	Slope	Elevation	Village	Type
Zone 1	+60-100	>20	>50	High	Nil+	Mangrove ground land or any type
Zone 2	+40-100	>20	35-50	Medium	Sparse/nil	Any type
Zone 3	+25-60	<20	35-50	Medium	Inhabitat	–do–
Zone 4	+<25	All	<35	Rolling	Habitat	–do–

The above procedures could well be achieved in the GIS domain. The study carried out in the Western Ghats showed the adequacy of the tool and approach. So, from the above Zone 4 and Zone 5 areas, it is possible to delineate the net areas available for JFPM activities and to indicate participatory forest management programmes. This should eventually restore the ecology and productivity of the area on a sustained basis.

8. Coastal vegetation and coral reef mapping

The coastal zones provide an ideal environment for growth and survival of coastal resources, such as mangroves and coral reefs, under certain environmental conditions. The mangroves are salt tolerant forest ecosystems that grow in intertidal regions. In the Indo-Pacific region, where their greatest diversity is found, they are unique in their great luxuriance. Several studies have been carried out using remote sensing in identification and mapping of mangroves in the middle Andamans, Sunderbans, Mahannadi delta and around the gulf of Kutchh. The protection of mangroves is important because they protect the coasts and harbours and are good breeding grounds for fish. They serve as firewood and fishing zones for the coastal fishermen and rural people.

Coral reefs are shallow water, tropical marine eco-systems known for their high rates of biological productivity. They are exceptionally diverse in fauna and have a complex food web and trophic organization. Satellite-based surveys have been carried out for the middle Andamans where, hard and living coral ecosystems have been identified using piece-wise linear stretch and the blue wavelength region.

9. Forest ground fire damage assessment

Of the 16 forest types in India, tropical moist deciduous forest is the most widespread, accounting for 37 per cent of the total forest cover (1981-1983). The other major type is tropical dry deciduous forest, which forms 28.6 per cent of the total forest cover. Generally, during summer periods, these forests are threatened by extensive man-made ground fires for regeneration of forest grasses for cattle. The fires occur primarily during the course of burning dry matter and grasses of the understorey, leading to damage to plantations and forest trees and preventing regeneration of saplings.

The use of satellite data permits foresters to estimate the ground fire damages periodically during the dry periods. The studies carried out in 1994 in Bandipur and Nagerhole national parks, in Simplipal, east India, have provided information on the spread of ground fires in January 1989, March 1989 and May 1989. The use of IRS-1C WiFS data in 1996 showed that it was possible to assess fire monitoring every five days in the dry deciduous forests of south India.

The use of multitemporal satellite data in the vulnerable areas would greatly help in making fire protection plans. A "Fire Alarm" model, integrating remote sensing with ground data, could be evolved for combating fire damage. The dry-season IRS-1C WiFS data has shown great promise for monitoring fire damage in dry deciduous forests in south India.

10. Shifting cultivation

The rapid degradation of tropical wet evergreen forests of the north-eastern region has become a growing concern due to the increased slash and burn activities by the local people and their shifting cultivation. Assuming a population growth rate of 20-30 per cent per decade, it is expected that 76,600,000 families will be dependent on jhuming by the turn of the century. Presently 4.35 million ha of land is under shifting cultivation. Several studies using remote sensing data have been carried out in parts of the north-eastern region.

There is a necessity to monitor the extent of degradation due to shifting cultivation and to provide compensatory rehabilitation areas for the local people to prevent further damage to forests.

11. Mapping and monitoring of river migration and loss of agriculture area

Every year, during the peak flow season, snow-fed rivers of the Himalayas cause devastating floods, heavy run-off and water flow resulting in erosion of river banks. This problem is more pronounced in the Brahmaputra and rivers flowing through the Indo-Gangetic plains. It is critical to study the annual migration pattern of the rivers, in order to account for the loss of land due to river bank erosion and creation of additional areas due to deposition. A study using remote sensing data, carried out on the Kosi River's migration, in Bihar, revealed that the river has migrated westward about 100 miles from its original course in the last two fifty years or more. This migration has caused extensive damage to alluvial plains and creation of waterlogged areas in the productive lands. Studies on river migration patterns would provide information for identification of the areas that are critically affected and enable people to take preventive measures.

12. Impact of mining on the environment

Intensive mining activity often leads to severe environmental degradation if timely and suitable corrective steps are not taken. Space images are now being regularly employed to assess the environmental damage resulting from extensive mining activities in various areas. As an example, the study carried out around Kudremukh iron ore mines in 1985 showed a significant decrease in the extent of forests and grasslands in this area, as a result of the mining activity, between 1976 (pre-mining phase) and 1985 (mining phase). The study revealed that about 358 ha was submerged by the Lakya tailing reservoir. The mines occupy about 116 ha, whereas plantations have been raised in about 138 ha.

Satellite data can also be used to map active and abandoned mines. Using satellite data of 1985, active and abandoned mines were mapped in Dehradun-Mussoorie limestone belt at 1:50,000 scale. In the same study it was possible to map the forests and the changes that have taken place during the periods 1973-1985. In another effort, methodology has been developed to identify the areas affected by underground coal fires in the Jharia-Raniganj coal fields using satellite and airborne thermal data.

13. Monitoring of coastline and estuarine environments

The geomorphic processes of erosion and deposition, along with sea level changes, modify the coastline. These effects are more pronounced in the estuaries where the tidal rise and fall affect the intermixing of fresh and saline waters.

(a) Mahi estuary

In a study carried out by the Space Applications Centre (SAC) in the Mahi estuary, using 100-year-old topographic maps, as well as satellite pictures of 1973, 1975, 1977, 1984, 1987 and 1988, the coastline changes over the area were revealed. The northern estuarine coast was under active erosion, particularly near the Dhuvaran thermal power station complex up to 1988. In the 1988 satellite data it was difficult to differentiate the power station's cooling pond and the river, whereas in 1973 a separation of about 5-7 km was observable from the coast. Another interesting feature seen in the mouth of the estuary is the development of shoals/islands through aggradation. This post-1975 development seems to be linked with the impounding of the river waters upstream in the Kadana and the Panam reservoirs, since flood waters which might have been flushing sediments from the estuary mouth into the sea are no longer available. The regulated flow apparently slows down the sedimentladen water as it enters the estuary and deposits its load. It is also likely that sediment brought into the estuary had caused the rapid erosion of the northern coast near Dhuvaran.

(b) Narmada estuary

Another study was done on erosion and deposition activities in Narmada estuary, where Aliabet Island is situated. Unlike the 100-year-old topographic maps depicting five islands/shoals in the estuary, the satellite images of 1988 show only one island other than the Aliabet. The size and shape of Aliabet Island has also changed drastically and it is growing in size. The river mouth has narrowed considerably on account of sediment deposition, while eroding the southern bank below Aliabet.

Thus the satellite data provides valuable information on coastal estuarine environment to monitor the processes of erosion and deposition of sediments.

14. Desertification

Desertification is a worldwide phenomenon affecting all the continents of the Earth and gradually decreasing the natural resources of land. Desertification is a process of diminution and destruction of the biological potential of the land, which leads ultimately to desert-like conditions, and is an aspect of the widespread deterioration of ecosystems under the combined pressure of adverse and fluctuating climate and excessive exploitation. As a result of man's impact on the ecosystems, mainly in arid and semi-arid regions, the extension of desert-like conditions or spreading of desert boundaries is common, involving complex physio-geographical processes that have negative consequences on land use, ultimately leading to a strong disturbance of the natural equilibrium. Factors include sparse vegetation, the pedological (soil) pattern and the surface water balance.

The immediate causes of desertification are well known -- overgrazing, the felling of trees for fuel, waterlogging, salinization and bad agricultural practices. Increasing population is also one of the causes.

The application of space-based images for monitoring the spread of deserts and the measures to arrest them is being attempted in India. The arid zone in India constitutes about 12 per cent of the geographical area, and the core tract lies in Ranjasthan, parts of Gujarat, Punjab and Haryana. The satellite remote sensing data has revealed that desert sands are drifting eastward through wind gaps (weak zones through which desert sands spread) in the Aravali ranges. There are 12 gaps located in the central and northern Aravali mountain regions, identified based on physiography, sparse vegetation, drainage pattern and other characteristics, as well as changes in the land use on the leeward side of the gaps. The study revealed that the coverage of affected area due to drifting of sand through these gaps is about 5,860 sq km, east of the Aravali ranges. Developmental programmes are being mounted for strengthening of these zones to arrest the further advance of the desert.

15. Identification and mapping of desert locust breeding centres

The desert locusts generally occur in two stages, recession and invasion. During the recession period, locust population is at low densities and infestation is confined to a 16 m sq km area in 30 countries from Atlantic Asia to north-west India. During the recession period the locusts live like ordinary grasshoppers, posing no problem. Desert locust upsurges that may develop into plagues are usually preceded by a period of widespread and heavy rainfall and subsequent development of seasonal vegetation in key breeding areas in successive seasons. The rain provides soil moisture for oviposition, egg development and growth of vegetation that provide food and shelter to the emerging hoppers. Rain also helps in rapid maturation of adults. Enormous damage can occur to agricultural crops during the plague periods, which are characterized by enormous swarms of hopper bands in many countries. This stream phase of invasion would cover normally 30 m sq km, which is about 20 per cent of the world's total land surface, covering 60 countries and affecting one-tenth of the world's population.

In India the desert locust is epidemic over 200,000 sq km spread over Rajasthan, Gujarat and Haryana states, wherein they breed and develop in monsoon periods. These areas are known as "Scheduled Desert Areas" of India. Many of the areas are inaccessible, and the sporadic rainfall, when it occurs, is often sufficient to induce locust breeding. Hence it is necessary to have periodic monitoring on a timely basis to identify the breeding centres that are inaccessible. The studies carried out by the Locust Warning Organization (LWO) using satellite data have made possible the identification of locust breeding centres, so that aerial spraying to arrest the further growth of locusts in these inaccessible areas can be done.

Since the locust menace causes extensive damage to agricultural production during the monsoon periods, the satellite remote sensing data would serve the purpose of identifying the breeding centres so that chemical measures by aerial spraying could be undertaken. This would also permit the government to keep the locust under perpetual recession solitary stage.

I. Conclusions

The tackling of environmental problems is intimately linked with many economic and social factors. The environmental consequences of human action also involve complex linkages among varied components of the Earth's environment, and there is still a greater need for better scientific understanding of the processes involved and the measures humankind should take to preserve the ecological balance while pursuing some of our urgent development needs. Space technology offers a powerful and speedy means for tackling some of the urgent and large-scale problems related to the environment, as exemplified by many applications realized already in India. Greater interdepartmental cooperation and sharing of knowledge and experience plays a crucial role in applying the use of technology, which shows great promise for assessing and combating environmental degradation and aiding sustainable development.

References

Government of India (1987). Biosphere reserves. In *Proceedings* of the National Symposium, Udhagamandalam, 24-26 September 1986.

The Hindu (1991). Science magazine section on water management policy, 12 November.

INAE (1990). *Water Management Perspectives, Problems and Policy Issues.*

Khoshoo, T.N. (1991). In *The Hindu* survey of the environment.

Madhav, Gadgil (1991). In *The Hindu* survey of the environment.

Mong Sunjoy (1991). The wasteland factor. *Sanctuary Asia*, 11(2).

Venkataramani, G. (1991). A man-made disaster. *The Hindu* survey of the environment.

PLANNING AND MANAGING REMOTE SENSING-BASED PROJECTS AT THE VILLAGE LEVEL: A CASE STUDY OF ANANTAPUR DISTRICT, ANDHRA PRADESH

*R.S. Rao**

ABSTRACT

The major and sustained adverse impact of drought is normally seen on agriculture, human beings and cattle population. The latest developments in space technology brought in remote sensing techniques and GIS to aid natural resource mapping and hazard management. The pilot integrated study presents a detailed micro-level plan at village level in order to more scientifically combat drought on a long-term basis using remote sensing techniques, which are time- and cost-effective. The basic information derived from remotely sensed data and conventional techniques about land and water resources at 1:50,000 scale have been integrated and Basic Integrated Land and Water Resource Units (BILWRU) have been mapped. Based on the BILWRUs, and keeping in view the present-day cropping patterns and the needs of the people, Recommended Optimal Land Use and Farming Systems (ROLUFS) were arrived at. Keeping in view the severe drought-situation and paucity of water resources, various drought-proofing works such as optimal in situ *soil and moisture conservation measures, such as rainwater harvesting structures, soil and moisture conservation measures, and fodder, fuelwood and permanent tree cover development zones, were recommended.*

The various scientific recommendations given are validated in the field through participatory rural appraisal (PRA) exercises. The district collector, who is the chairman of the District Rural Development Agency, dovetails the developmental funds available under various programmes. The programme is implemented through a watershed development committee comprising the beneficiaries themselves with the assistance of non-governmental organizations (NGOs). Large-scale participation of NGOs in a campaign mode has led to a massive people's movement for watershed development. Early results obtained by implementation of the recommendations are presented briefly. The role of satellite remote sensing for monitoring these programmes in the form of normalized differential vegetation index (NDVI) before and after implementation is also highlighted.

A. Introduction

Anantapur District, in the south-western corner of Andhra Pradesh State, covers an area of 19,135 sq km with a population of 3,183,814. The population density is 166 persons per square kilometre against state average of 242 persons/sq km. The district has 964 villages falling in 63 *mandals* (administrative divisions). Out of the total population, 76.5 per cent are in rural areas; 15.4 per cent are cultivators 17 per cent constitute agricultural labourers and 54 per cent are non-workers. Out of the cultivators, 41 per cent are small farmers, 49 per cent are marginal farmers and only 10 per cent are large farmers. The district has only 42 per cent literacy rate, and the decennial population growth rate is 24.95 per cent.

Anantapur forms part of the northern extremity of the Bangalore plateau and slopes from south to north. The district lying off the coast does not enjoy the full benefit of the north-east monsoon,

* Remote Sensing Applications Centre, Hyderabad, India.

and, being cut off by the high Western Ghats, the rainfall from the south-west monsoon is also prevented, so the district is deprived of both the monsoons and is subject to recurring drought. The average precipitation in the district is 516 mm per annum, spread over four seasons. The rainfall decreases from east to west from about 700 mm to about 250 mm. The rainfall is not only scanty but erratic in nature. There were severe droughts in the years of 1896-1897, 1900, 1901, 1924, 1925, 1934-1935 and 1941-1942. The analysis of rainfall data for the last 100 years reveals that out of the last 10 decades, seven have been drought-affected decades, and in the last 10 years, six years were rainfall-deficient years resulting in severe drought conditions. The district is devoid of significant perennial surface water resources and hence only 7.89 per cent of the geographical area is irrigated, which is largely dependent on ground water. Nearly 55 per cent of the irrigated area is under wells. The district is predominantly occupied by hard rocks, hence the ground water potential is also assessed to be limited. The terrain is undulating, resulting in a high run-off and contributing to active erosion of the top layer of the soil. The evapotranspiration is also high due to the semi-arid climate, high temperature and high wind velocity.

The net sown area is about 69 per cent of the geographical area in spite of the drought conditions, since the bulk of the population is dependent on agriculture. The major crop is groundnut, which is cultivated under rainfed conditions. The average yield of the crop is 689 kg per hectare, against the state average of 867 kg/ha. After the crop is harvested the land is left fallow and that results in absence of work avenues for the landless poor and agriculture labourers.

The perpetual occurrence of drought has left the district with vast stretches of barren rocky areas or degraded soils with very poor crop yields. Soil erosion has become a major problem, which in turn is silting up innumerable irrigation tanks. Thus the meager surface water storage available is also decreasing day by day. Sustaining rainfed agriculture is also becoming a Herculean task. The decreasing surface water resource availability has kept the pressure on groundwater, and over-exploitation of the limited resource has become a common practice. This has resulted in fall of water levels, so most of the dug wells that tap the weathered and fractured part of the hard rock (to an average maximum depth of 20-25 m) have gone dry. The forest areas, which were supporting elephants in 1900, are devoid of any major vegetation cover and have been reduced to scrub.

B. Integrated Mission for Sustainable Development (IMSD)

The Department of Space has sponsored a project titled Integrated Mission for Sustainable Development (IMSD), co-sponsored by the government of Andhra Pradesh. The state Remote Sensing Applications Centre (APSRAC) has carried out the study specifically to find scientific and lasting solutions to recurring droughts at the micro-level, i.e. in villages.

IMSD aims at generating natural resource information and preparing action plans for development of the terrain on a sustainable basis, in order to keep up the productivity and maintain the quality of the environment. A detailed database on natural resources, terrain conditions, socio-economic status and demography is a prerequisite to preparing such action plans. Keeping in view the concept of sustainability, it is particularly important that the databases be presented in the form of maps to facilitate spatial analysis.

The methodology of the study involves generation of thematic maps showing current land use/land cover, types of wastelands, forest cover/types, surface water resources, drainage pattern, potential groundwater zones, land-forms (geomorphology), geology (rock types, structural features, mineral occurrence) and soil types, using data from Indian Remote Sensing satellites. A map showing slope/aspect is prepared using topographic contour information, and the meteorological data (rainfall intensity, distribution and so on) are collected from existing databases. By integrating the above thematic layers using GIS techniques, derivative maps called Basic Integrated Land and Water Resource Units (BILWRU), showing resource availability of individual land parcels, land characteristics, status of soil erosion, priority watersheds needing immediate treatment and other features, are prepared. These maps are validated through adequate field checks and by using authentic existing information on various aspects. Socio-cultural, socio-economic and demographic information is gathered from the existing databases and through selective field surveys. Specific developmental plans for these units are arrived at in consultation with and close coordination between space scientists, experts from various central/state developmental departments, agricultural universities/research institutions, district-level officials and local farmers and villagers, so as to ensure the technical feasibility and cultural acceptability of such action plans.

Based on the BILWRUs, and keeping in view the present-day cropping patterns and the needs of the people, Recommended Optimal Land Use and Farming Systems (ROLUFS) are arrived at in consultation with various scientists of agricultural institutions. Keeping in view the severe drought situation and paucity of water resources, site-specific drought-proofing works such as optimal *in situ* soil and moisture conservation measures, regarding rain water harvesting structures, soil and moisture conservation measures, fodder, fuel wood and permanent tree cover development zones, are recommended.

C. Natural resources of Anantapur District

Detailed natural resource mapping at 1:50,000 scale has been taken up using satellite remote sensing data of IRS-lA satellite (plate 1a). Information like rainfall, socio-economic data and so on that is not amenable to remote sensing is collected through conventional techniques.

Rainfall: The rainfall of the district has been analysed for the last 100 years, and the average rainfall is observed to be higher than the current period's average annual rainfall, taking into account the very good rainfall periods of the early part of the century. Hence, the rainfall pattern of the last 30 years has been analysed and it is observed that there is a steady decline in the rainfall trend (figure 1). It is very clear from figure 1 that out of the last 34 years, 17 years are lower than normal. There are years such as 1984, 1985 and 1986, in which there is a failure of the monsoon. However, it is also observed that the district has received very good rainfall, almost touching 800 mm, during some years. In 1989, the district saw unprecedented rainfall and run-off which resulted in flash floods, breaching 37 irrigation tanks. Taking the clue of availability of good rainfall, the IMSD programme has been designed to harvest the rainfall of the good rainfall years and store it below the ground so that the same can be used during the periods of drought.

The weekly rainfall analysis for 1994 (figure 2) shows that a substantial part of the rainy season was receiving much lower than normal rainfall of the week, as low as 10 mm. There are severe dry spells such as in 13th, 14th, 15th and 16th weeks of 1994, which is the crucial period for crop production. The district annual average rainfall works out to 516 mm, which means a total precipitation of about 9,950 million cu m.

Surface water: In the district, there are three major rivers, the Penner, Hagari and Chitravati, flowing through the district. They are mostly dry except for seasonal run-off. There are about 3000 irrigation tanks which are able to hold only about 8 per cent of the precipitation, i.e. 840 million cu m.

Geology: The geological formations in the district are a group of metamorphic rocks belonging to the Archaean age; they consist of schists, gneisses, migmatites and younger granites, pegmatites, quartz veins and basic dykes. Archaean rocks suffered a considerable degree of tectonic disturbances, as a result of which, the rocks have been metamorphosed and recrystallized. Alluvium can be seen along the river courses of the Penner, Hagari, Chinna Hagari and Chitravati rivers.

Hydro-geomorphology: The major portion of the district forms a pediplain carved out of Archaeall gneisses, schists and granites with hill ranges and hillocks of relatively small relief scattered all over. Based on geomorphic expression, geology, relief, soil cover, structure and groundwater prospects, the district has been divided into different hydro-geomorphic units. Valley fills, flood plains and moderately weathered pediplains along the fractures are found to be groundwater potential zones. The well inventory data collected during finalization of the hydro-geomorphological maps provided a better appreciation of the relation between groundwater potential and geomorphic units. Groundwater in the predominantly hard rock area of the district occurs in the weathered and fractured rocks under the water table and in semi-confined to confined conditions. With the introduction of fracturing, and because of weathering, they have developed secondary porosity, which has improved the chances of tapping better yields and more than often have given rise to potential aquifers. The degree of weathering varies from less than a metre near the outcrops and hill slopes to more than 20 m in the valley bottoms and topographic lows. This weathered zone has been tapped extensively by dug wells, dug wells-cum-barillas, and barillas which invariably tap the fractures occurring below the weathered zone. The yields vary from less than 20 cu m/day to 90 cu m/day.

Status of groundwater development: Groundwater resource potential estimation at village level for the entire district has been carried out. The inflow components considered are recharge from rainfall, recharge from applied irrigation and tank seepage. The outflow components taken are irrigation draft and domestic draft. Finally, the groundwater balance and the status of development were arrived at for each of the villages, covering the entire district. Thus the villages with overdraft conditions are identified.

Artificial groundwater recharge: The major source of recharge of groundwater is rainfall. On assessment of water resources of Anantapur District, it is observed that out of the total rainfall received, only 11 per cent is going down into the groundwater, and about 8 per cent is stored as surface water, while about 60 per cent is lost in evapotranspiration and about 20 per cent goes out as surface water run-off. To ensure that at least 20-30 per cent of the rainwater is used for recharging the groundwater, rainwater harvesting and artificial groundwater recharge structures need to be constructed in the entire district. These water harvesting structures help in storing the rainwater and allow it to percolate down to the groundwater, thus raising the groundwater level.

Based on the status of groundwater development and predominant groundwater-irrigated areas, artificial recharge structures such as check dams, percolation tanks, farm ponds and subsurface dykes were recommended to recharge the wells in the downstream areas to augment the groundwater supply during drought spells.

Rural drinking water supply and fluorosis problem: Groundwater is a major source of drinking water in rural villages of Anantapur District. The high fluoride level of more than 1.5 ppm is observed to be injurious to health. There are about 210 villages in Anantapur District affected with dental and skeletal fluorosis. The problem was endemic, and the communities had no answers to it. However, remote sensing-based selection of sites for drilling barillas in about 87 villages to locate low-fluoride groundwater sources was recommended and implemented with 80 per cent success.

The occurrence of high fluoride-content groundwater in the district is quite well known because of its association with the rock types such as younger granites occurring in the district, which have high fluoride-bearing apatite and fluorapatite minerals. The areas with prominent high fluoride-bearing waters of more than 1.5 mg/l occur mostly in Kadiri, Penukonda, Dharmavaram, Anantapur, Gorantla, Gooty, Rayadurg, Kalyanadurg, Ramgiri, Tadpatri, Narpala and Singanamala. Village-wise groundwater samples were analysed for fluoride measurements. The analysis shows the fluoride-rich groundwater is being diluted by the low-fluoride surface waters downstream of the surface water body as a result of augmented infiltration. In view of these observations, rainwater harvesting structures such as check dams and percolation tanks are suggested upstream of the low fluoride zone so that the impounded water would not only recharge groundwater but also help in diluting fluoride-rich groundwater, thus serving the dual purpose of both quality and quantity. Therefore, 200 check dams have been located to dilute the drinking water sources.

A major drinking water supply scheme was taken up by Sri Satay Sai Trust for providing drinking water to 790 villages at a cost of 170,000 million rupees in a record time of eight months. The scheme, which started on 1 February 1995, was scheduled to be completed by 23 November 1995. Remote sensing-based hydro-geomorphological maps, followed up by ground hydrogeological and geophysical surveys, were used to select drilling sites for 190 villages. In a record time of two months, all of the site selection and drilling was completed (by 31 March) with a success rate of more than 90 per cent.

Soils: The soil series in the district have been mapped using remote sensing and field information, and 96 series have been identified. The classification of soils according to soil taxonomy places these soils into the orders of entisols, inceptisols, vertisols and alfisols. These have been further divided into sub-groups in each of the orders, which have a number of soil series under them. The soils in the very deep layers are very dark grey to black, clayey and highly calcareous; in the deep to moderately deep layers, they are light brown to yellowish red, gravelly, sandy clay loam; and in moderately deep to shallow layers, the soils are dark brown, clay loam, and slightly calcareons. Soil salinity and alkalinity development and sheet erosion are aoso predominant the soils in general are poor in nutrient status.

Land use/land cover: The district has various land-use/land-cover classes, which are derived from multi-date satellite data. The district is predominantly occupied by kharif unirrigated areas; irrigated areas using surface water and groundwater resources can be seen along the valley portions. The other land-use/land-cover classes include built-up lands, kharif irrigated, rabi irrigated, double crops, fallows, degraded underutilized forest, barren rocky stone areas, salt-affected lands, land with or without scrub, and sheet rock areas. The net sown area is about 69 per cent of the geographical area.

Forests cover one tenth of the total district, while barren and uncultivable lands like hills and others, cover 8.74 per cent. The area put to non-agricultural use, such as buildings, roads and waterways, covers about 6.3 per cent, while there is a perceptible increase in the net area sown. The only persistent drawback is the vast extent of barren and uncultivable land. About one fifth of the area is arable, but fallow. A single dry crop is raised under rainfed conditions both in black and red soils. In the black cotton soils, late kharif or early rabi crop is taken to facilitate tilling in black soils.

Slope: Slope and altitude are very important from the land utilization point of view. The maps showing slope aspect are prepared using Survey of India topographic maps at 1:50,000 scale.

Socio-economic status: The primary activity in the district is agriculture. Industrial development is comparatively insignificant in this district. With respect to the social situation, social justice and equality are the two important aspects to be considered. With regards to inequalities in the social sector, the land holdings are particularly uneven in their distribution, except in a few pockets which are primarily dry areas. Based on the socio-economic data supplied by the district administration, which were gained by conventional surveys, maps were prepared showing demographic details, size of farming community, number of farmers living below poverty line, fodder and fuelwood demand and availability, biogas plants, and drinking water needs. The infrastructure status was also analysed based on standard score method, and *mandals* with different infrastructure facilities are evaluated.

1. Integration and recommendations

The data obtained by using remote sensing techniques as well as conventional methods were integrated using a geographic information system (GIS) approach to prepare derivative maps called Basic Integrated Land and Water Resources Units or BILWRUs. Recommendations were made by APSRAC for each of the BILWRUs, based on the water, soil, slope, land use and socio-economic aspects such as recommended, optimal land use and farming systems (plate 1d). The various recommendations given include:

(a) Intensive agriculture with drip irrigation, in areas with good groundwater potential, good soil and gentle slopes;

(b) Irrigated dry crops are recommended in areas with moderate groundwater prospects and gentle to medium slopes on good soils; horticulture is recommended in moderately good groundwater zones with good soils;

(c) Dry farming is recommended in areas with poor groundwater potential and moderately good soils, with soil and moisture conservation measures.

The various crops suitable in each BILWRU were identified and recommended.

2. Recommended drought-proofing works

Various drought-proofing works are recommended to reduce the impact of drought (plate 1c). The *mandal*-wise details of recommended drought-proofing works are given in table 1. They include:

(a) Rainwater harvesting structures such as percolation tanks, check dams, farm ponds, diversion drains, and subsurface dykes are recommended at specific sites;

(b) *In situ* soil and moisture conservation includes vegetative barriers, contour bunding with stone checks and other measures; soil erosion control measures include planting of soil-binding species, gully control works, broad bed and furrow method of cultivation, conservation ditches, irrigation water management and horticultural species planting on field bunds;

(c) Fodder, fuelwood and forest development includes fodder development in tank foreshores and marginal agricultural lands, reclamation of salt-affected lands and planting salt tolerant/resistant fodder species, fuelwood plantations, reclamation of salt affected lands, planting salt tolerant/resistant fuelwood species, social forestry and silvipasture, casuarina and soil-binding species, afforestation/silvipasture, afforestation with contour trenches, quarrying with environmental protection measures, and avenue plantation along all roads and railway tracks.

3. Rainwater harvesting structures

Based on geology, geomorphology, soils and drainage, rainwater harvesting structures such as checkdams, mini-percolation tanks and farm ponds are recommended upstream of the groundwater irrigated areas to replenish the aquifer immediately. The areas of dykes, reefs, outcrops, lithologic contacts, black soil areas, saline soils and tank commands, on the other hand, should be avoided.

Check dams are recommended upstream of groundwater irrigated areas on lower order streams, whereas mini-percolation tanks are recommended on second- or third-order streams. The catchment area of the check dams is usually less than the catchment area of mini-percolation tanks, and the capacity of each structure is less than 1 million cu ft. In all, 2,400 rainwater harvesting structures, mostly percolation tanks and check dams are recommended, out of which the district administration has already constructed 1,600. Most of the structures received six to eight fillings during the monsoon of 1993 and 1994.

Farm ponds are recommended for capturing run-off from the agricultural fields to have one irrigation for the dryland agriculture. They need not be located on the streams but they can be located on sites wherein 15-20 ha of catchment can be tapped.

They are recommended more in black soil areas than in red soil areas because run-off is better assured. Surface water resources in this area can be exclusively allocated for artificial recharge through a network of recharge structures after duly desilting existing surface water bodies. This will result in greater irrigation potential, while solving the rural water supply problems, as apposed to the conventional practice, in which surface water resources are directly used for irrigation. This can be

Table 1. *Mandal*-wise details for recommended drought-proofing works in Anantapur District

Mandal	No. of rainwater harvesting structures				Area in hectare for *in situ* soil and moisture conservation				Area in hectares for fodder, fuelwood and forest development										Road length in km for avenue plantation
	1	2	3	4	5	6	7	8	9	10	11	12	13	14	15	16	17	18	19
Anantapur	670	28	13	--	15,573	33	1,765	4,146	--	945	1,084	721	68	362	--	--	730	55	273
Raptadu	616	10	5	--	12,233	463	170	5,651	88	--	--	382	447	805	--	--	--	70	243
Garladinne	340	27	14	--	11,206	56	4,358	4,541	--	802	230	1,226	126	--	--	579	1,575	5	250
Atmakur	531	34	66	--	14,071	33	83	5,077	160	--	680	631	20	--	--	442	2,727	315	248
Kuderu	900	82	55	--	20,207	894	60	4,693	40	2,427	572	268	--	--	--	1,380	1,063	748	327
Singanamala	220	--	12	1,000	9,195	150	1,145	8,591	--	3,212	522	1,167	--	--	--	496	5,374	--	258
B. Samudram	323	13	8	--	8,732	195	1,960	9,430	2	1,095	398	1,044	--	--	--	180	--	8	233
Narpala	418	30	39	--	7,128	15	220	7,485	38	3,490	112	1,573	--	--	--	1,876	1,073	13	263
Tadipatri	674	11	7	--	9,970	--	14,171	4,173	--	4,587	165	--	--	--	175	38	135	--	440
Yadiki	455	1	--	--	1,821	--	7,506	2,996	--	2,302	--	188	142	4,165	--	138	4,796	--	316
Peddapappur	422	10	3	--	219	--	4,911	3,172	--	1,002	359	1,190	--	--	--	1,731	--	--	305
Putlur	668	9	1	--	145	--	9,812	427	--	3,807	10	--	--	--	--	2,251	2,666	--	336
Yellanur	374	20	4	--	1,795	40	5,693	1,740	77	2,275	--	1,126	--	--	--	1,670	5,890	5	301
Guntakal	647	9	35	2,250	15,924	190	3,621	4,554	10	2,415	505	45	--	--	--	448	3,686	399	355
Gooty	204	4	3	--	5,258	--	45	4,825	--	2,550	1,487	404	--	--	--	--	1,232	--	312
Pamidi	256	3	1	--	7,068	23	150	7,166	53	430	661	615	--	--	--	1,430	2,076	--	183
Peddavadugur	388	15	--	--	5,324	--	485	3,234	--	1,317	940	1,009	--	--	--	240	2,498	--	356
Uravakonda	902	21	21	--	14,028	799	3,835	3,547	--	--	--	1,047	125	--	--	--	--	198	393
Vajrakarur	1,104	49	4	1,500	10,853	35	19,071	2,092	--	--	106	1,955	--	--	--	1,633	618	208	504
Vidupanakal	927	1	--	--	2,617	2,969	24,187	--	--	75	93	148	--	680	--	20	--	--	425
Dharmavaram	411	30	8	1,750	18,778	610	124	6,492	103	--	297	2,220	986	--	--	173	2,783	185	413
Tadimarri	573	36	13	3,000	2,753	--	--	5,495	41	1,319	774	1,102	--	--	--	15	35	165	232
Battalapalli	383	4	4	1,500	12,946	--	--	6,315	20	172	233	592	172	--	--	8	305	30	282
C.K.Palli	408	68	44	8,000	15,635	658	71	5,253	170	4,978	960	936	--	--	--	113	2,810	91	362
Kanaganapalli	631	48	23	1,000	24,798	188	385	4,848	139	9	227	725	512	3,274	--	1,420	1,849	25	444
Ramagiri	533	17	13	4,000	15,197	398	783	2,897	88	1,420	103	233	404	3,949	--	865	873	45	281
Kalyandurg	483	33	17	--	33,162	68	--	11,685	65	792	140	229	--	--	--	360	1,422	115	522
Beluguppa	179	15	9	--	11,963	10	--	6,963	137	509	107	94	--	--	105	180	460	28	400
Kambadur	679	32	18	--	22,491	713	--	8,354	146	471	477	803	--	--	--	290	88	169	323
Kundurpi	329	30	35	--	15,907	1,174	--	8,893	18	1,064	836	486	--	--	--	328	1,783	5	292
Brahmasamudram	198	13	7	--	18,531	50	--	5,669	78	110	660	610	--	--	--	230	413	78	386
Settur	388	11	2	--	20,469	213	--	5,465	85	1,053	1,106	585	--	--	--	320	240	--	327
Rayadurg	283	44	10	2,500	18,669	115	--	1,346	50	120	173	1,961	--	--	--	280	2,740	63	367
D. Hirehal	360	18	6	--	12,582	--	--	4,461	--	772	320	34	--	739	--	600	2,753	15	416
Gummagatta	292	3	3	--	16,278	188	--	5,257	--	--	35	363	--	274	--	223	688	90	267
Kanekal	503	--	--	--	6,642	--	--	4,447	--	813	81	139	--	--	295	--	--	--	459
Bommanhal	214	--	--	--	5,607	--	--	2,957	--	--	301	323	70	1,129	113	50	363	--	336
Penukonda	332	33	21	500	8,554	903	--	4,549	139	6,665	584	1,192	--	--	--	464	6,463	77	242

Table 1 (Continued)

Mandal	No. of rainwater harvesting structures				Area in hectare for *in situ* soil and moisture conservation				Area in hectares for fodder, fuelwood and forest development										Road length in km for avenue plantation
	1	2	3	4	5	6	7	8	9	10	11	12	13	14	15	16	17	18	19
Somandepalli	380	26	16	500	8,517	560	--	4,934	163	1,530	80	813	333	1,581	--	133	888	24	206
Roddam	655	--	3	--	17,523	379	1,513	7,665	586	2,634	905	994	--	--	--	--	45	36	302
Puttaparti	139	13	10	--	8,766	80	--	2,631	--	3,624	32	1,268	--	--	--	2,240	4,677	137	158
Kothacheruvu	170	8	4	750	8,454	179	--	10,394	104	--	--	1,075	240	--	--	22	2,802	86	218
Bukkapatnam	167	5	--	1,250	7,496	38	--	2,450	--	1,621	50	1,738	--	--	--	64	1,007	92	185
Madakasira	656	43	12	1,000	20,863	10	--	4,980	210	--	--	1,410	655	--	--	595	2,048	--	349
Amarapuram	421	7	15	--	12,308	--	--	7,518	547	--	210	663	257	--	--	--	--	253	250
Gudibanda	318	43	22	500	13,437	--	--	4,692	331	--	408	723	--	--	--	633	1,000	28	229
Rolla	217	22	32	--	8,787	73	--	3,070	50	344	32	817	--	--	--	1,008	547	70	134
Agali	111	--	2	--	7,036	--	--	2,367	214	823	78	884	36	--	--	--	--	50	136
Hindupur	203	2	--	--	8,283	30	13	6,473	71	--	--	1,168	325	--	--	167	650	--	236
Parigi	210	7	5	--	4,756	48	--	7,130	--	--	--	72	68	--	--	--	--	--	172
Lepakshi	242	15	11	2,250	6,994	20	8	2,839	43	207	192	434	--	--	--	181	312	--	178
Chilamatur	486	14	6	--	11,780	258	--	4,973	15	1,008	79	1,094	--	--	--	409	691	278	216
Gorantla	627	53	6	--	16,875	347	--	5,941	77	1,430	291	1,580	--	--	--	384	2,336	114	234
Kadiri	319	14	11	1,500	10,817	233	--	3,020	--	--	--	958	20	--	--	110	1,048	14	199
Mudigubba	849	6	--	6,250	30,544	319	--	5,680	81	2,969	177	1,543	--	--	--	85	5,131	304	412
Nallamada	340	13	44	4,250	12,499	127	--	5,680	--	--	--	1,065	30	--	--	243	1,120	164	241
N. Pulikunta	877	20	--	--	11,220	256	--	1,771	--	5,622	--	1,013	--	--	--	811	5,157	60	306
Talupula	562	21	20	275	16,134	861	--	3,693	--	3,070	23	1,398	--	--	--	102	12	43	246
Nallacheruvu	315	10	8	2,000	9,623	98	--	2,600	--	251	--	1,220	12	--	--	50	--	62	96
O.D. Cheruvu	507	36	28	1,000	13,231	--	--	4,363	--	1,236	70	1,772	--	--	--	230	1,425	64	228
Tanakal	766	21	19	--	9,639	663	--	4,785	--	3,909	39	1,336	208	--	--	28	939	43	202
Amadagur	463	23	2	500	10,197	279	--	3,005	--	1,176	--	1,071	--	--	--	517	4,946	191	111
Gandlapenta	440	37	8	750	8,845	53	--	1,795	--	1,214	--	1,010	18	--	--	997	15,473	31	189
Totals	28,658	1,271	808	49,775	748,953	16,094	106,145	303,335	4,239	85,666	18,004	54,485	5,274	16,958	688	29,480	114,461	5,349	18,105

Notes:
1. Farm ponds.
2. Check dams.
3. Percolation tanks.
4. Diversion drains.
5. Vegetative barriers, contour bunding with stone checks, etc.
6. Soil erosion control measures including planting of soil binding species, gully control works etc.
7. Broad bed and furrow method of cultivation and conservation ditches.
8. Irrigation water management and horticultural species planting on field bunds.
9. Fodder/plantation development in tank foreshores.
10. Fodder development in marginal agricultural lands.
11. Reclamation of salt-affected lands and planting salt tolerant/resistant fodder species.
12. Fuelwood plantation in marginal lands.
13. Reclamation of salt-affected lands and planting salt-tolerant/resistant fuelwood species.
14. Social forestry and silvipasture.
15. Casuarina and soil binding species.
16. Afforestation/silvipasture.
17. Afforestation with contour trenches.
18. Quarrying with environmental protection measures.
19. Avenue plantation along all the major roads.

achieved owing to the judicious irrigation water management under wells benefiting from artificial recharge, minimizing the losses due to evaporation and unproductive seepage.

4. Implementation

The various scientific recommendations given are validated in the field through participatory rural appraisal (PRA) exercises. The rural population of a village is assembled for a day-long discussion and field visits with the district officials of various sectors. The inputs from space technology are presented to the people and their reactions to the suggestions are discussed at length. A model of their village is drawn by the people with coloured chalk powder locally known as *rangoli.* Standing around this coloured model, the scientists and the people debate on various pros and cons of individual recommendations given vis-à-vis the traditional practices and traditional wisdom. Thus an operational, implementable and people-acceptable programme is finalized.

The District Collector, who is the chairman of the District Rural Development Agency, dovetails the developmental funds available under various programmes like Drought Prone Area Programme, Employment Assurance Scheme, Integrated Wasteland Development Programme, Special Component Plan, Tribal Sub-plan, afforestation programmes of the Forest Department, Minor Irrigation Programme, National Watershed Development Programme, soil conservation programmes, animal husbandry department programmes, and others. The programme is implemented through a watershed development committee comprising the beneficiaries themselves with the assistance of non-government organizations (NGOs). The watershed development committee is composed of a president, secretary and beneficiary members with powers to operate a joint bank account wherein the watershed development funds are deposited in instalments. The beneficiary groups of the Development of Women and Children in Rural Area (DWCRA) programme and their family members form the nucleus of these watershed development committees; thus the self-help groups already mobilized in a village are drawn to participate in the watershed development programme. The women who are active in the DWCRA programme influence their husbands or male members of their family to take up the various drought-proofing works under watershed development. This leads to a massive participation of the rural farmers, agricultural labourers and people below the poverty line, especially those belonging to weaker sections. The non-governmental organizations/voluntary organizations who are active in the district are also encouraged to mobilize the people's participation and form beneficiary groups. Large-scale participation of NGOs in a campaign mode has led to a massive people's movement for watershed development. Thus the watershed development teams participate physically in actual implementation of the programmes of the watershed development and also contribute either labour or logistics or (if the benefit goes to the entire village) funds available from the village panchayats such as Jawahar Rojgar Yojanat and Intensive Jawahar Rojgar Yojana. A district-level committee was formed with participation from non-governmental organizations, experts as well as direct beneficiaries. Financial resources from various programmes, such as the Prime Minister Employment Assurance Scheme (EAS), Drought-prone Area Programme (DPAP), Jawahar Rojgar Yojana (JRY), Intensive Jawahar Rojgar Yojana (IJRY) and Schedule Caste Action Programme (SCAP), were dovetailed.

Water conservation, with the help of check dams, has been an important feature of the implementation work. Check dams are low-priced cement structures, recommended along stream courses in medium slope areas, upstream of groundwater irrigated areas to recharge groundwater and act as water points for cattle.

Accordingly, the district administration has taken up the construction of rainwater harvesting structures, and a total of about 3,000 check dams and percolation tanks have been completed. The groundwater levels and the extent of area under irrigation before and after the rainwater harvesting structures were monitored.

D. Results of implementation in Vanjuvanka watershed

The ecological problem common to the entire district is more pronounced in this watershed. The area receives just about 300 mm of rainfall annually, the lowest being about 250 mm and highest being 400 mm. In the last 10 to 15 years, before the implementation of the IMSD project, there has been consistent drought, and only about 25-40 per cent of the crop is successful. Because of delayed and erratic monsoons, people had not been able to sow on time or reap the groundnut crop.

As per IMSD recommendations, groundnut is intercropped with red gram, which is a deep-rooted crop. This not only helps in soil regeneration, but also gives additional yields to the farmer. Along with the intercropping pattern, another recommended practice, that of ploughing along the slope and contour bunds with a trench inside, has been implemented. This helps in arresting the water velocity, reduces soil erosion and helps in moisture retention. Construction of rainwater harvesting structures has been undertaken on recommended sites. About 57 check dams and mini-percolation tanks have been built along with diversion drains and farm ponds.

Rockfill dams are interspersed with the check dams to prevent soil erosion and siltation. These rockfill dams help in slowing down the flow of water. Desiltation of tanks is being taken up in conjunction with application of silt on the fields to improve the soil fertility and moisture-holding capacity.

Vegetative barriers and contour bunding with stone checks have been built to prevent soil erosion and siltation of water harnessing structures. Apart from this, stone checks have been made to treat the gulleys that have formed over the years.

To provide fodder and fuelwood and to restore natural forest cover, alternate land-use systems are being implemented in the marginal lands, tank foreshores, saline lands and degraded forest land. Areas with gentle slopes in forests have been brought under silvipasture. Contour trenches and contour bunding with afforestation is carried out on steep sloping forest areas. Planting of soil-binding species like the casuarina have been taken up to arrest migrating river sands. Human and cattle trespassing has been checked by cordon walls around the afforested hills. Local utilitarian species of trees such as neem, tamarind, pongamia, babul and others have been planted in different nurseries. In one central nursery sponsored under the EAS scheme and managed by an NGO, avenue plantations have been undertaken along all the road and railway tracks. Such activities have helped restore the natural forest cover, which is urgently required.

A detailed analysis of the normalized differential vegetation index (NDVI) of the watershed before and after implementation has shown an increase of about 48.3 per cent. Here, in the Vanjuvanka watershed, slowly but surely the valley is giving nature a wide berth to thrive upon.

There is no doubt that things have changed for the better in Anantapur District. More than anything else, the difference has been made possible due to the support and participation of the people at the grass-root level. Unlike most projects, where the beneficiaries form a mute audience, the farmers participating in this project are quite vocal.

Rich mulberry plantation for sericulture, and other cash crops like sunflower, are grown in the watershed. The catchment areas of check dams, the tank foreshore, the farm bunds, the contour trenches, the saline lands and the hilltops all support vegetation. Horticulture in form of banana plantations, papaya plantations, sethapnal plantations, coconut trees and others can also be seen. In the three to four years of operationalization, the impact is found in terms of soil and water conservation, groundwater recharge and irrigation possibilities being developed in the watershed.

An impact analysis was also carried out to see the effect of the rainwater harvesting structures. The analysis clearly indicated a rise in the water table and an increase in the command area under wells located downstream of the structures, indicating augmented recharge from the percolation tanks/

check dams (table 2), which ultimately results in sustainable development of the watershed. Groundwater levels were monitored at 37 wells. The effect of filling in these structures was significant within three or four days of rainfall, and in about 15 days the groundwater levels stabilized.

Table 2. Effect of check dams on the groundwater level in Vanjuvanka watershed, Anantapur District, Andhra Pradesh

Name of the village	No. of wells benefited	Groundwater extraction structure	Total depth in m	Water level (bgl) before/after in m	Rise in water level
Karutlapalli	4	DW	7.0	Dry/3.9	>3.1
		DW	11.4	Dry/7.5	>3.9
		DW	15.0	Dry/10.0	>5.0
		DW	11.0	Dry/8.1	>2.9
Karutlapalli	2	DW	13.0	Dry/12.1	>0.9
		DW	14.0	Dry/10.6	>3.4
Karutlapalli	2	DW	12.6	10.6/7.5	3.1
		DW	--	9.0/7.8	2.8
Kadadarakunta	3	DW	9.6	Dry/9.0	>0.6
		DW	10.8	Dry/10.3	>0.5
		DW	10.8	Dry/10.3	>0.5
P. Yaleru	1	BW	--	28.7/18.0	10.7

The cropping pattern has transformed the entire landscape from brown to green.

Resource limitations, both human and financial, plague most of our well-intended and well-planned projects. This limitation has been taken care of in the IMSD approach. There are various NGOs who are committed to working in the area of sustainable development and have already made progress in the field.

Apart from the NGO activities, integration of various other government programmes -- such as the DWCRA programme for the women and children or the TRYSEM programme for self employment among youth -- provided a ready-made network leading to a cross-section of people in rural areas. Ultimately the operationalization is achieved through campaigns in the village involving the youth, the DWCRA group members, the voluntary organizations, the Gram Saksharta Samiti, and others.

The women involved in economic activities under the DWCRA, in a carpet-weaving centre, are also sensitized to other issues related to health care and nutrition, family planning and environment protection. Such places are much more than money-earning centres.

E. Conclusion

Every effort for sustainable development in the district is channelled towards IMSD activities, and the achievements so far indicate that work on other micro-watersheds along the same lines is likely to meet with the same success. As in the case of any successful experiment, the Anantapur approach too presents a number of dimensions and perspectives for viewing the process of comprehensive and sustainable development.

In recent years, attempts have been made to regenerate Anantapur's natural resources and bring a semblance of normalcy to its disturbed ecosystem. A lot of barren lands today carry lush vegetation. In many places the depleting groundwater table has been checked. Soil that can support healthy vegetation has been provided due protection. And along with all this, there are people sensitized to the environmental problems and aware of the need to lend their help in bringing Anantapur back on its feet with the help of space technology at the village level.

Now, in these watersheds, rainwater harvesting through check dams, percolation tanks, farm ponds, contour bunding across the slopes, vegetative barriers, horticulture, avenue plantation, afforestation with silvipasture, fodder in foreshore areas of tanks, fuel wood plantations, and other methods has become a common sight. Extinguishing of forest fires by people themselves has become a common practice. The integrated development of watershed programmes has led to substantial water availability, which in turn has generated additional irrigated crops; the general vegetation cover over the area can now be seen in the form of increased vegetation.

Based on the encouraging results of the pilot studies carried out in various districts of the country, locale-specific action plans that integrated thematic information on land and water resources were derived from remotely sensed data, meteorological data and socio-economic information; then, a nationwide project entitled Integrated Mission for Sustainable Development (IMSD) was launched in 172 problem districts of the country. These districts, covering nearly 45 per cent of India's geographical area, are those perennially affected by drought and flood; they also come under hilly and tribal areas and are under active study for implementation of India's development programmes.

Bibliography

Rao, R.S., M. Venkataswamy, C. Mastan Rao and G.V.A. Rama Krishna (1993). Identification of over-developed zones of groundwater and the location of rainwater harvesting structures using an integrated remote sensing based approach: a case study in part of the Anantapur District, Andhra Pradesh, India. *International Journal of Remote Sensing,* 14(17): 3,231-3,237.

Rao, R.S., M. Venkataswamy, G.V.A. Rama Krishna and C. Mastan Rao (1994). Role of remote sensing in integrated fluorosis control programme in Anantapur District, Andhra Pradesh, India. In *Proceedings of International Space Year Conference on Remote Sensing and GIS* (ICORG'94), pp. 309-314.

Rao, R.S., M. Venkataswamy, G.V.A. Rama Krishna and C. Mastan Rao (1994). Integrated remote sensing-based rainwater harvesting programme and its impact in Vanjuvanka watershed, Anantapur District, Andhra Pradesh, India. In *Proceedings of International Space Year Conference on Remote Sensing and GIS* (ICORG'94), pp. 622-627.

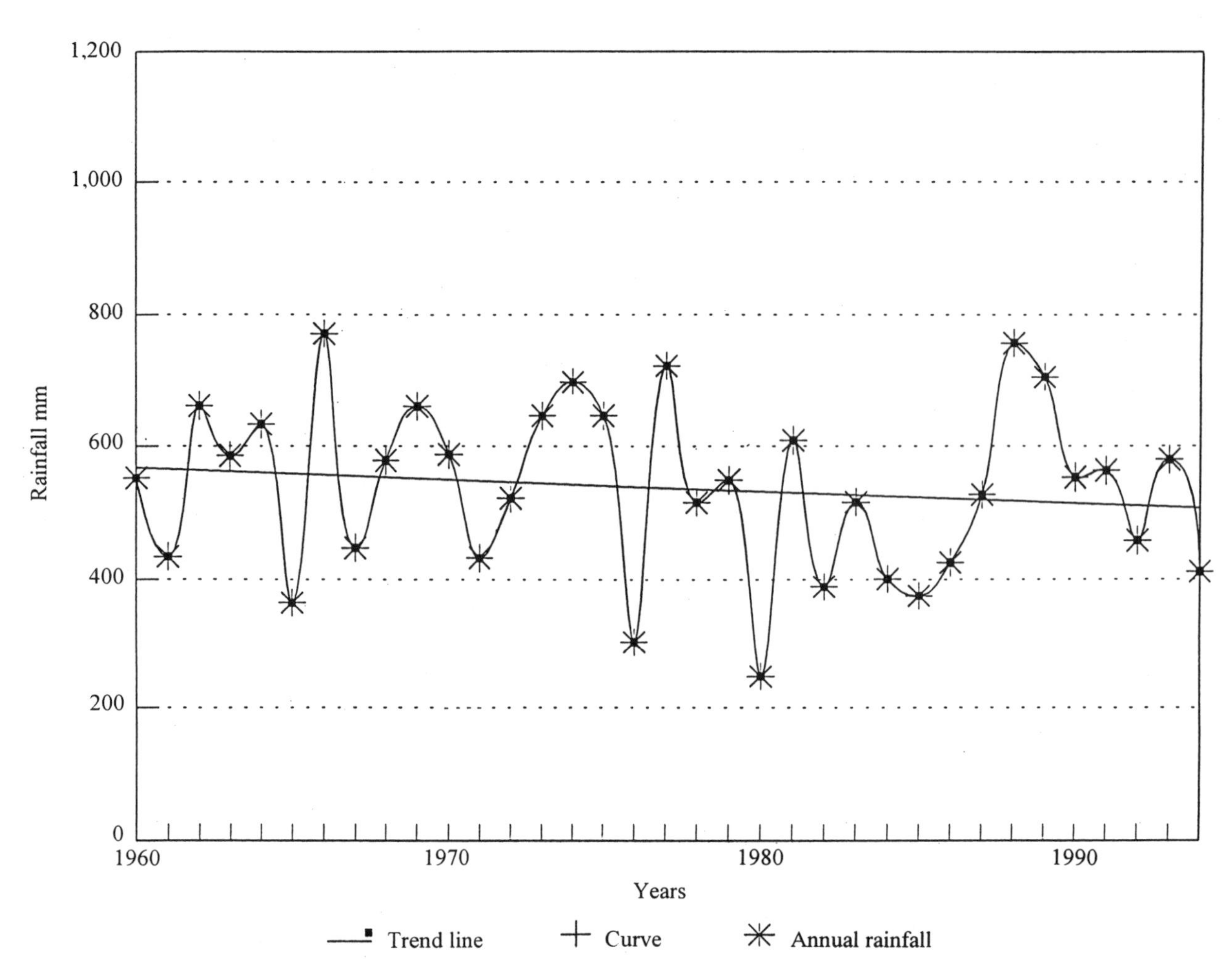

Figure 1. Rainfall in Anantapur

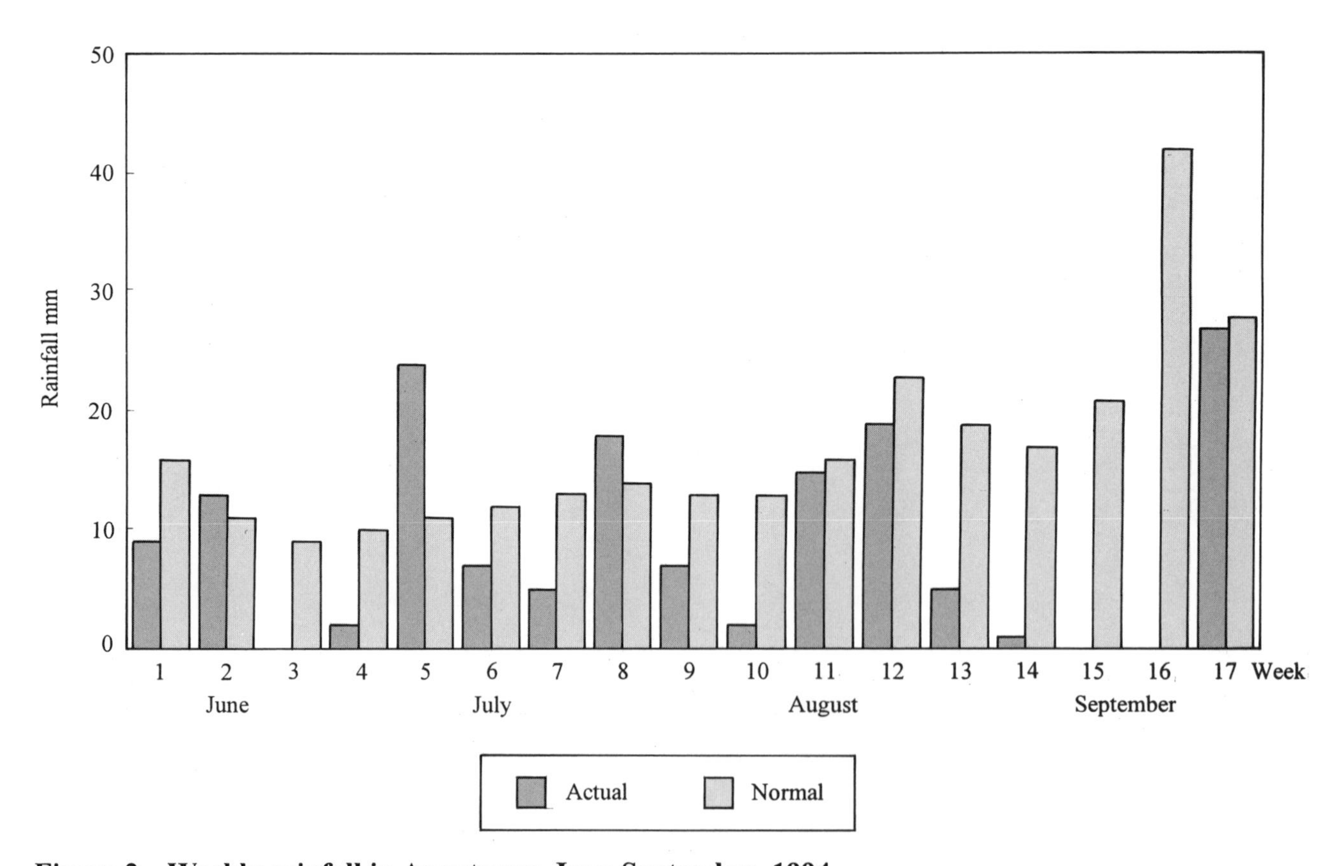

Figure 2. Weekly rainfall in Anantapur, June-September, 1994

PART FOUR

LAND USE/LAND COVER

LAND USE/LAND COVER MAPPING USING SATELLITE DATA: THE MALAYSIAN SCENARIO

K.F. Loh **and** *S. Ahmad**

ABSTRACT

Malaysia has participated in several bilateral and regional projects in land use/land cover mapping using satellite technology. The experience gained from these projects has aided the operationalization of land use/cover mapping in the country on a large scale, using optical satellite data. The major shortcoming encountered is the occurrence of persistent and extensive cloud cover in certain areas, making it difficult for updating land use/cover information on a regular basis. The use of SAR data, although still at a "juvenile" stage, has met with some measurable success in alleviating the above problem. The Malaysian scenario over the last fifteen years or so in land use/cover mapping using satellite data is presented in this article.

A. Introduction

The availability of satellite imagery in different forms has attracted a flood of applications in rural development planning because of its cost effectiveness, near real-time information acquisition and digital nature, which enables sophisticated data manipulation and analysis in a computer. Malaysia has operationalized the use of satellite data for land use/cover mapping and continuous forest monitoring. Other applications, which include agro-ecology zoning and spatial planning for sustainable agriculture production using the integration of remote sensing and GIS, have been conducted on a pilot basis.

Since 1989, with the establishment of the Malaysian Centre For Remote Sensing (MACRES), Malaysia has participated in several bilateral and regional projects in the utilization of satellite data to expedite the acquisition of land use/land cover information.

Under the Malaysia-Sweden cooperation efforts between 1990 and 1993, a practical classification scheme for land use/cover mapping using satellite data in the Malaysian environmental setting was developed. The scheme is being adopted by the Department of Agriculture in producing the national land-use map, using primarily Landsat TM data.

The major shortcoming in the use of optical satellite data for land use mapping is the occurrence of persistent and extensive cloud cover over certain areas of Malaysia, making it difficult to update information on a regular basis. SAR data with its ability to penetrate clouds has helped alleviate this shortcoming with some measurable success.

Malaysia's participation in the EC-ASEAN ERS-1 and Japan JERS-1 verification programmes indicated that SAR data alone are of limited value in land use mapping despite their advantage over optical data in the detection of some land-use types. Composite images comprising varied combinations of optical and SAR data provide slight improvements over optical data in land use mapping. However, temporal SAR data has considerable potential for effective change detection and, subsequently, updating of existing land use/cover baseline information derived from optical data.

This paper highlights the Malaysian scenario in the utilization of satellite data for land use mapping -- the past, present and future.

* Malaysian Centre for Remote Sensing, Kuala Lumpur.

B. Malaysia-Sweden Cooperation Project

This project was implemented jointly by MACRES and the Swedish Space Corporation (SSC) for a three-year period starting from 1990. One of the objectives of the project was to effect a transfer of technology on land use/cover mapping using satellite data from Swedish experts to Malaysian counterparts. The project area consisted of the states of Kedah and Selangor (figure 1), covering 9,500 and 8,200 km^2, respectively.

An important output of the project was the development of a classification scheme suitable for land use/cover mapping in Malaysia based on monoscopic visual interpretation of satellite data. The scheme was used for the production of 18 land-use/cover map sheets over Kedah and 15 over Selangor at 1:50,000 scale. Altogether, 26 SPOT multispectral scenes were used in the mapping. Landsat TM data were also used to complement SPOT in areas where problems of overlapping spectral signatures of cover types were encountered. This was particularly true among mature stands of forest, oil palm and rubber located adjacent to one another, which could not be visually differentiated on the SPOT images.

Ideally, land-use and land-cover classification schemes should be prepared separately, but in practice the two are often mixed, especially when satellite remote sensing data are the principal mapping sources. The land-use/cover classification scheme developed under the Malaysia-Sweden project consists of three levels, comprising 23 mappable units at 1:50,000 scale, as appears in table 1. The scheme is similar to that used for land use mapping using aerial photography in Malaysia (table 2), except that the mappable units have been reduced from 30 to 23.

Figure 2 shows a mosaicked and precision-corrected SPOT image of two scenes corresponding to map sheet 3,658 of the Kuala Selangor area.

Table 1. Classification scheme for land use/land cover mapping using satellite data

Level 1	Level 2	Level 3
Urban land	–	Urban and associated area (1U) Industrial area (1Ui) Estate settlement (1E) Recreational area (1A) Mining area (1T)
Cultivated land	Horticulture	Mixed horticulture (2H)
	Tree crops	Rubber (3G) Oil palm (3O) Coconut (3C) Forest plantation (3F)
	Non-tree crops	Paddy (4P) Sugar cane (4Y) Diversified crop (4C)
Non-cultivated and vegetated land	Open land	Grassland (6R) Bush (6B)
	Forest land	Dryland forest (7F) Coastal swamp forest (7Fc) Logged-over forest (7L) Secondary forest (7S)
Non-cultivated and non-vegetated land	– –	Cleared land (9F) Natural bare land (9N)
Water		Large water body (10W) Large fish-pond (10K)

Table 2. Classification scheme for land use/land cover mapping using aerial photographs

Level 1	Level 2
Settlements and associated non-agricultural areas	Urban and associated areas (1U) Estate buildings and associated areas (1E) Tin mining areas (1T) Other mining areas (1X)
Horticultural lands	Mixed horticulture (2H) Market gardening (2M) Agricultural stations (2E)
Trees, palm and other permanent crops	Rubber (3G) Oil palm (3O) Coconut (3C) Pineapple (3N) Tea (3T) Coffee (3K) Cocoa (3A) Pepper (3P) Sugar cane (3Y) Orchards (3X) Sago (3S) Banana (3B)
Cropland	Paddy (4P) Diversified crops (4C) Shifting cultivation (4X) Tobacco (4T)
Permanent pasture	Permanent pasture (5)
Grassland	Grassland (6)
Forest land	Forest (7F) Scrub (7S) Newly cleared land (7C)
Swamps, marshlands and wetland forests	Swamps, marshlands and wetland forests (8)
Unused land	Unused land (9)

The successful implementation of the project has, since 1995, resulted in the operationalization of land use/cover mapping using satellite data on a large scale in Malaysia. The Malaysian experience shows that there is little difference in the use of SPOT or Landsat TM for land-use mapping, but because of its better spectral resolution and cheaper data source, the Landsat TM is preferred. The Department of Agriculture is presently using Landsat TM to conduct land-use surveys and mapping of the whole country, using the classification scheme adopted under the Malaysia-Swedish project.

More and more baseline land-use/cover information, visually derived from satellite data, are being digitized in a geographic information system (GIS). Updating of this database has been conducted from time to time by overlaying the vector digital land-use/cover file on the satellite image in the GIS; polygon boundary updating is normally done by screen digitizing.

The use of satellite data has expedited land use/cover mapping sixfold compared to the conventional aerial photography mapping. Mapping with aerial photos was started in the 1950s by the Department of Agriculture, during which time land-use/cover maps covering the whole country were produced every seven years. With the use of satellite technology, the country expects to reduce the duration to two years.

C. EC-ASEAN ERS-1 and JERS-1 verification projects

Persistent and extensive cloud cover over certain areas of Malaysia has limited the use of optical satellite data for reliable land use/cover mapping. Malaysia has used SAR data to complement the optical data to resolve this problem and has met with reasonable success.

Under both the EC-ASEAN ERS-1 (1993-1996) and JERS-1 verification (1995-1996) projects, Malaysia has experienced at first hand the complementary nature of SAR and optical data for land use/cover mapping. The conclusions arrived at are as follows:

(a) Backscatter and image brightness analysis indicate that SAR images have better visual differentiation for land-cover types such as bananas, senescent oil palm and built-up areas than optical images have. However, this does not imply any significant advantage in favour of SAR data for land use/cover classification. In fact, multitemporal SAR data by themselves are of limited value in land use/cover mapping because of small differences in the backscatter of most cover types;

(b) Composite images with varied combinations of SAR and optical data provide only marginal improvement in visual differentiation over the composites of optical data;

(c) There is considerable potential in the use of multitemporal SAR images for change detection and, subsequently, the updating of baseline maps prepared from optical data.

Figure 3 shows a composite of Landsat TM 453 of the Kuala Muda, Kedah, area, and figure 4 is a composite of JERS-1 SAR and TM bands 4 and 3 over the same area. The latter image provides a better differentiation of cover types in the urban and paddy areas. In figure 5 the land-use/cover vector file derived from the Landsat TM data (a) was overlaid on the multitemporal SAR data (b) for the purpose of updating changes in the forest and oil palm covers that have occurred between 1991 and 1994.

The Department of Agriculture is presently updating the land-use/cover database derived from both aerial photos and Landsat TM data from change detection analysis, using temporal SAR data in areas severely affected by cloud cover.

D. The future

Land-use/cover mapping is an ongoing process in Malaysia. Given the rapid development Malaysia is experiencing, the country will continue to use satellite data for this purpose to expedite acquisition of timely land-use/cover information for decision makers. The classification scheme developed under the Malaysia-Sweden project will be modified from time to time to cater for forthcoming higher resolution satellite data and more advanced computer processing techniques.

Since Malaysia is in the tropics, where cloud cover is persistent and extensive, the focus on the utilization of SAR to complement optical data for land-use/cover mapping, so updating becomes inevitable. It is also expected that more and more SAR missions, equipped with state-of-the-art SAR technology (hyper-frequency, interferometry and polarimetry), will be launched in the future. With the current advancement of SAR technology, a time will come when SAR data can be completely independent of optical data for land-use/cover mapping, a development that will be extremely useful for tropical countries like Malaysia.

E. Conclusion

Satellite technology has played and will continue to play a major role in land-use/cover mapping in Malaysia. The country's experience in using satellite data for land-use/cover mapping shows that optical data like SPOT and Landsat TM are useful, but in areas of severe cloud problems it should be complemented with SAR data. Multitemporal SAR data has considerable potential for change detection and for subsequent updating of land-use/land-cover maps.

Bibliography

R.E. Brown, M.G. Wooding, A.J. Batts, K.F. Loh and K.M.N. Ku Ramli (1995). Complementary use of SAR and optical data for land cover mapping in Johore, Malaysia. ERS-1 Application Workshop, London, 6-8 September.

K.F. Loh (1994). Land use/cover mapping of the state of Selangor, Malaysia, using satellite data. Decision Makers Seminar on Application of Remote Sensing and Geoinformation System, Langkawi, Malaysia, 12-16 December.

K.F. Loh (1995). Land-use/cover mapping of the state of Kedah, Malaysia, using satellite data. International Symposium on Vegetation Monitoring, Chibai University, Japan, 29-30 August.

K.F. Loh, K.M.N. Ku Ramli and L. Nordin (1995). Complementary nature of SAR and optical data for land cover mapping. Conference on Remote Sensing and GIS for Environmental Resources Management, Jakarta, Indonesia, 6-8 June.

K.F. Loh, K. Jusoff, I. Selamat, Z.A. Hassan and S. Ahmad (1996). Potentiality of SAR temporal tata for land-use/cover change detection and map updating. National Conference on Climate Change, Universiti Pertanian, Malaysia, 12-13 August.

N.N. Mahmood and K.F. Loh (1990). Economics of implementing the Malaysian Natural Resources Evaluation programme using remote sensing technologies. ESCAP/ADP Regional Conference on the Assessment of the Economics of Remote Sensing Applications to Natural Resources and Environmental Development Projects, Guangzhou, China, 14-18 November.

N.N. Mahmood, K.F. Loh, I. Selamat and Z.A. Hassan (1996). Updating of land-cover maps using ERS-1 SAR data. Seminar on Technology for Updating Maps Using Remote Sensing, Jakarta, Indonesia, 22-24 July.

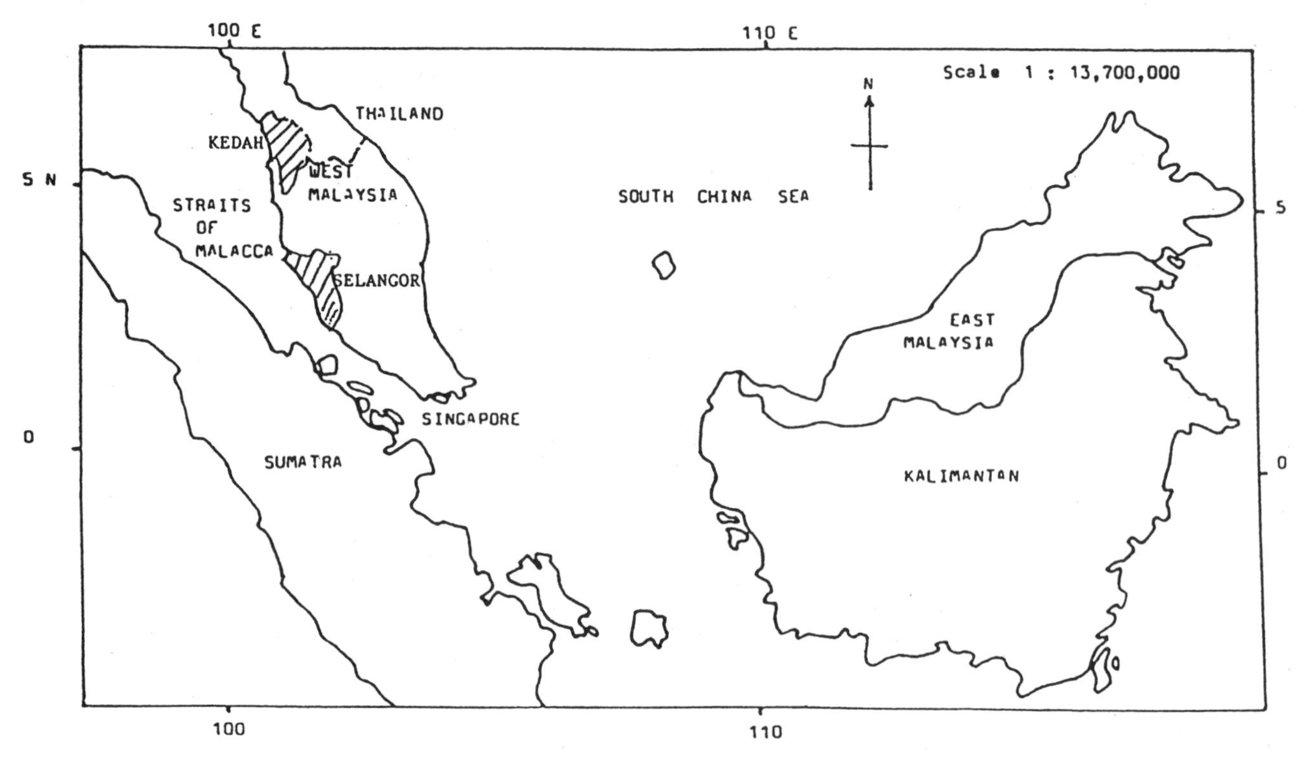

Figure 1. Location map of the states of Selangor and Kedah, Malaysia

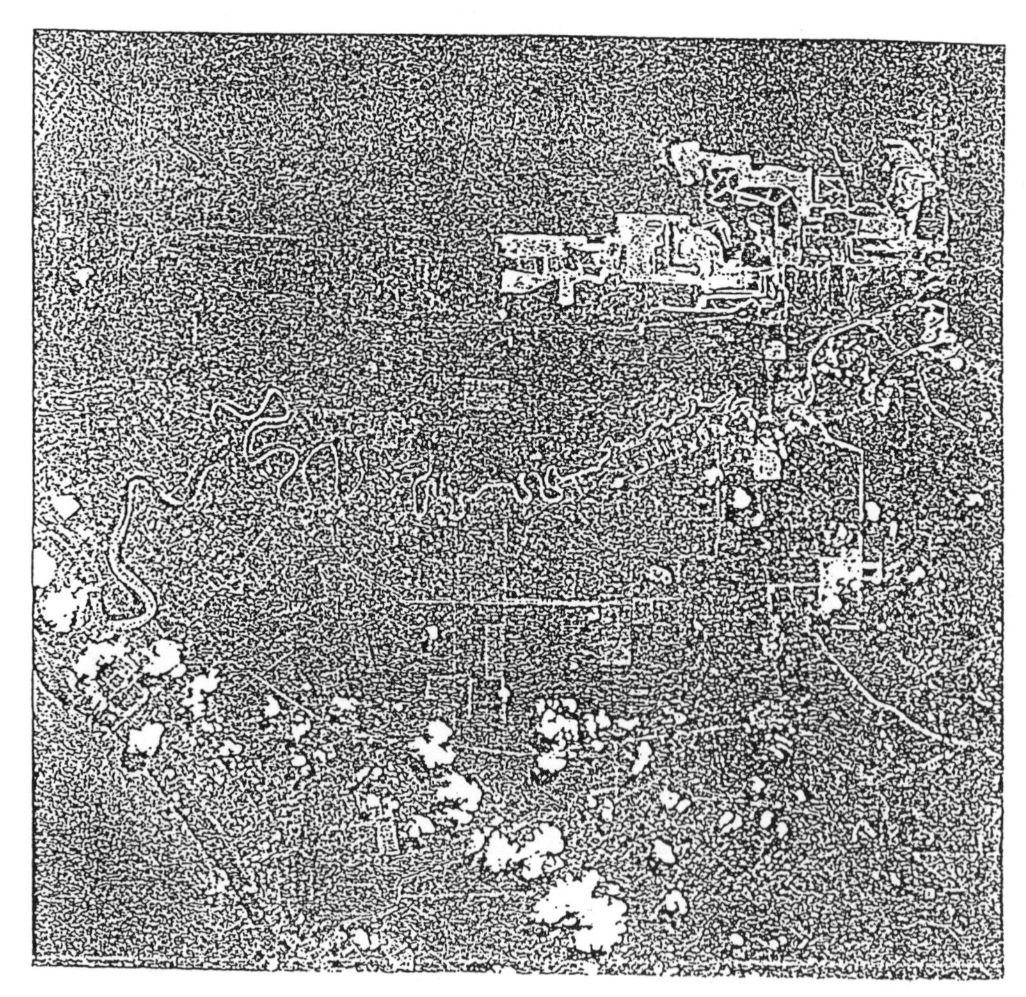

Figure 2. Mosaicked and precision-corrected SPOT image (2 scenes) of Kuala Selangor area

Kota Kuala Muda, Kedah

Landsat TM -- 20/2/1994
Band 4: Red; Band 5: Green; Band 3: Blue

Figure 3. Landsat TM (453) composite image of Kuala Muda, Kedah

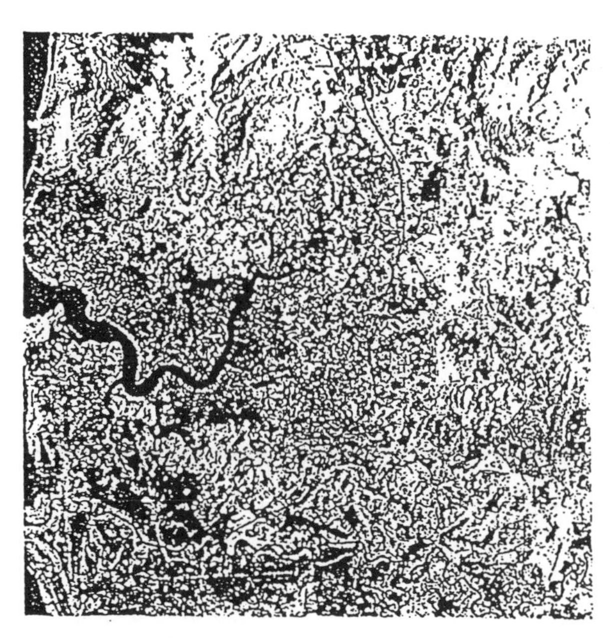

Kota Kuala Muda, Kedah

JERS-1 SAR	– 13/4/1994:	Red
Landsat TM Band 4	– 20/2/1994:	Green
Landsat TM Band 3	– 20/2/1994:	Blue

Figure 4. Composite image of JERS-1 and TM bands 4 and 3 of Kuala Muda, Kedah

Landsat TM image, 11 March 1991
with 1990 land-use map.
Red: Band 4; Green: Band 5; Blue: Band 3.

_____ Forest cleared to oil palm on TM image, but not yet updated on land-cover map.

Figure 5(a). Landsat TM 453 (1991) with vector land-use overlay

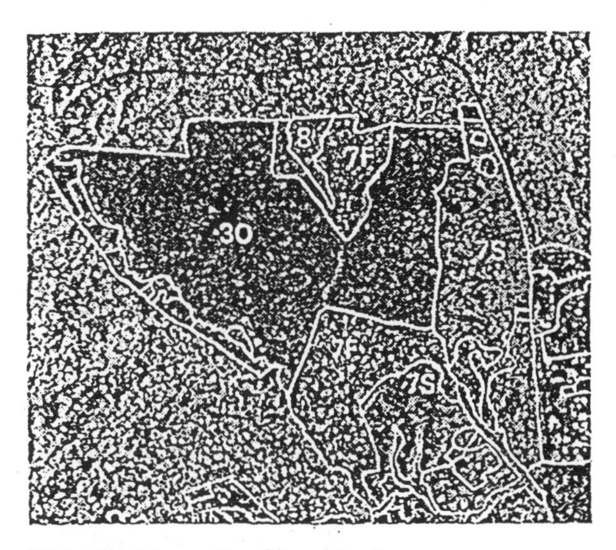

ERS-1 SAR image, Lee filtered 3 x 3,
with updated land-use map.
Red: 4 January; Green: 25 January; Blue: 18 February 1994.

Land-use codes

30: Oil Palm 7F: Forest
7S: Scrub 8: Wetland Forest

Figure 5(b). ERS-1 temporal composite (1994) with vector land-use overlay

REMOTE SENSING APPLICATION FOR LAND USE MANAGEMENT IN INDIA

*N.C. Gautam, R. Nagaraja and V. Ranghavswamy**

A. Introduction

The advent of satellite remote sensing has brought in a phase change in the information systems of the world. It has had different impacts in different parts of the world. In those countries where the information system was of a high order of accuracy, the question asked was how much more the remote sensing could improve accuracy. In such situations the remote sensing might not have contributed much in improving the accuracy but it made a significant contribution in making the information more timely. In developing countries where no or obsolete information systems were in vogue, remote sensing had a significant impact on both aspects: improving the accuracy as well as making the information near real-time. The land-use management systems in developing countries are riddled with problems of a socio-economic nature on one hand and lack of an information system on the other. Remote sensing has proved to be a very effective means of developing an integrated information system that can meet the challenges of managing lands for meaningful and sustainable utilization.

B. The Indian scenario

Although the central government in Indian exercises control on formulation and regulation of national policies on land use planning, the executive responsibility for managing land lies with the individual states within the country. There is a common hierarchy in all states which dictates administrative justification on the basis of functional considerations. This may vary in size from the entire state to smaller units. The district administrative unit, on the basis of revenue returns, is the unit for planning and execution of management activities in the country.

The land use data collected by village administrations, using eye estimation, will be compiled at *taluk* (district), state and finally at the national level to form the basis for planning. However, these data lack the spatial dimension and do not show the status of land transformation over a period of time; that is to say, maps are needed.

The management of land has to involve problems of major concern: the encroachment on forests by other land uses such as agriculture and homesteads; the impact of multi-purpose river valley projects, mining; land degradation due to soil erosion and mismanagement of water; shifting cultivation; and encroachment on agricultural land and forest land to meet the urban requirements of the civilization. The rapidly shrinking forest cover warrants increasing afforestation efforts, and the greening of wastelands must also be undertaken.

C. Remote sensing applications for land use management in India

Land use and land cover information is the basic prerequisite for the conservation and management of land, water and vegetation resources. The information on land use/land cover available today in the form of thematic maps and published statistical figures in records and publications is inadequate, inconsistent and does not provide up-to-date information on the changing land-use patterns, processes and their spatial distribution in space and time. Efforts made by the state and central government departments and other institutions to bridge the information gaps and remove inconsistency in ground data collection, data reporting and data compilation procedures, though encour-

* National Remote Sensing Agency, Balanagar, Hyderabad, India.

aging, is slow and time-consuming. It is here where satellite remote sensing offers an alternative, accurate and fast mode of data collection and updating of the land use/land cover information, enabling authorities to arrive at a standard classification and explanation of different land-use/land-cover classes.

Realizing the need for up-to-date and accurate land-use/land-cover maps by several departments in the country, especially for agricultural land use planning in the 15 agro-climatic zones, the National Remote Sensing Agency (NRSA), in the Department of Space (DOS), has carried out land use/land cover mapping, at the behest of the Planning Commission. The mapping uses a hybrid methodology for agro-climatic zone planning, covering all 442 districts in the country at 1:250,000 scale using IRS (LISS-I) satellite data of 1988-1989 for two cropping seasons, kharif and rabi. A land-use/land-cover classification system comprising twenty-two classes (up to Level II) was developed by NRSA in 1989.

D. Definition of land use/land cover

"Land use" can be defined as "man's activities and the various uses which are carried out on land".

"Land cover" refers to "natural vegetation, water bodies, rock/soil, artificial cover and others noticed on the land" (NRSA, 1989).

Since both land use and land cover are closely related and are not mutually exclusive, they are interchangeable terms, since the former can be inferred based on the land cover and on the contextual evidence.

E. Methodology

The methodology followed for developing the classification system is described hereunder.

1. Visual interpretation

The methodology developed by National Remote Sensing Agency for visual interpretation of the multitude of satellite data is composed of the following six major steps (see also figure 1):

(1) Selection and acquisition of data: Standard FCC imagery of IRS LISS-I for kharif and rabi seasons;

(2) Preliminary visual interpretation of data: IRS LISS-I FCC imagery of kharif and rabi seasons are interpreted individually making use of the interpretation keys. The boundaries of land-use/land-cover classes are plotted onto a transparent overlay, such as artian or polyester tracing sheets;

(3) Ground data collection and verification: Following the previously drawn scheme and traverse plan, ground truth information is collected following specific pro forma procedures to cover at least 10 per cent of the district area. Areas of doubtful preliminary interpretation are particularly verified;

(4) Final interpretation and modification: Based on the ground truth data, modifications are effected and classes as well as their boundaries refined;

(5) Area estimation: Areas under different classes are estimated by planimetric measurements to complete district land use statistics;

(6) Final cartographic map preparation and reproduction: Fair drawing originals are made as per pre-designed specifications and cartographic symbols.

Land-use/land-cover maps of 273 districts in the country have been completed following the visual interpretation approach.

2. Digital classification

The operation methodology of land use mapping was not available in the digital domain, particularly for large area coverage involving two seasons of data. The methodology for digital classification, thus developed, comprises the following seven steps (see also figure 2):

(1) Data acquisition, loading, merging and geo-referencing;
(2) Ground truth collection;
(3) Demarcation of district boundaries and transfer of administrative and cultural features;
(4) Stratified classification of the data from the two seasons;
(5) Refinement;
(6) Aggregation of rabi and kharif classifications;
(7) Statistics and final output.

Supervised classification is carried out in a stratified manner for both rabi and kharif seasons with maximum likelihood classifier. Stratification is effected by digitization of forest boundaries from Survey of India (SOI) topographical maps (also other sources) and creation of mask files. The classification proceeds through selection of the training sets, calculation of the statistics of the training sets and the decision boundary of maximum probability based on mean vector, variance, co-variance and correlation matrix of the pixels.

The normalized difference vegetation index (NDVI) is adopted for classification of forest areas. Gaussian post-normalization is used to bring the NDVI within the range of 0-255. A look-up table is prepared for output grey values for each project area, based on limited ground truth.

Through a specifically-developed program, REFINE, the methodology allows human logic and intuition to correct and modify discrepant classifications arising out of spectral similarity of two or more classes in a non-forest stratum. In referential refinement, classification of a pixel in one cropping season is evaluated and, if necessary, corrected with reference to its classification in the other season.

Rabi and kharif classified scenes are reformed into a single output through the process of pixel-to-pixel aggregation. Finally, land-use statistics and the photo-write compatible tape of the aggregated classified output are generated for the project area.

This methodology has been successfully adopted in generating district-wise land-use/land-cover maps of 168 districts.

3. Results

It is seen from the area under broad (Level I) land use/land cover, the agriculture area is 165.24 million ha or 50.26 per cent; wastelands 75.53 million ha or 22.98 per cent; forests 47.62 million ha or 14.40 per cent (excluding areas under degraded forest and forest blanks, which amount to 18.08 million ha or 5.50 per cent); water bodies 10.60 million ha or 3.22 per cent; and built-up area 13.91 million ha or 4.24 per cent (percentages are in reference to the total geographic area of the country). Other areas, which include salt-pans and settlements mixed with vegetation areas, account for 7.37 million ha or 2.28 per cent of the total geographical area. The detailed (Level II) land-use/land-cover classes of 22 classes and their area statistics are shown in table 1.

4. Comparison of statistics with ground-surveyed area figures

The land-use/land-cover area statistics derived from satellite data at the district level for the base year 1988-1989 were compared with the figures published by the Bureau of Economics and Statistics (BES) and the Ministry of Agriculture (MOA) for the same period, to check for consistency and compatibility. Firstly, the comparison was made by matching the different land-use classes in both to "one-to-one" common groups, and, secondly, the definitions/explanations of each class were compared with respect to the "nine-fold classification" of land utilization adopted by BES and DES/MOA. On comparison, differences in area were noticed with respect to the "net area sown"

Table 1. Area under detailed land-use/land-cover categories, India (1988-1989)

Category	Area in ha	Percentage of total geographic area
Built-up land	13,913,772	04.24
Agricultural land		
Kharif land	120,586,769	
Rabi crop land	76,300,445	
Double-cropped area	53,105,792	
Net area sown ($)	151,481,769	46.09
Fallow land	13,762,590	04.17
Forest		
Evergreen/semi-evergreen	14,184,366	04.32
Deciduous forest	31,813,875	09.68
Degraded forest	16,274,270	04.95
Forest blank	1,813,853	00.55
Forest plantations	1,119,452	00.34
Mangroves	504,999	00.15
Wasteland		
Salt-affected	1,988,380	00.60
Waterlogged	1,219,666	00.37
Marshy/swampy	823,876	00.25
Gullied/ravined	2,020,329	00.62
Land with or without scrub	26,514,564	08.06
Sandy area (coastal and desertic)	5,572,086	01.69
Barren rocky/stony waste/sheet rock area	6,251,414	01.91
Water bodies		
River/stream	8,414,852	02.55
Lake/reservoir/tank/canal	2,195,968	00.67
Others		
Grassland/grazing land	3,104,538	00.94
Mining/industrial waste	116,497	00.03
Shifting cultivation	2,823,626	00.87
Snow-covered/glacial area	6,992,255	02.14
Salt pans	37,808	00.01
Unclassified	7,452,095	02.27
Unsurveyed (*)	8,329,400	02.53
Total	328,726,300	100.00

Source: Report on area statistics of land use/land cover generated using remote sensing techniques, 1995, NRSA, Hyderabad.

Note: ($) Includes 7,700,343 ha or 02.34 per cent of area also under agricultural plantations.
(*) Unsurveyed area is in Jammu and Kashmir due to hanging boundaries.

(NAS), "land put to non-agriculture use" and "pasture land". But the major difference in areas was observed with respect to "net area sown" and "land put to non-agricultural use".

The difference in the area reported under NAS was around 15-20 per cent less than the figure derived by remote sensing in most of the districts and states in the country. To enable researchers to understand the reasons for differences in the figures, the Planning Commission and the Ministry of Agriculture requested a reconciliation survey (verification of information derived from remote sensing techniques and the BES ground data collection/reporting system). The survey was jointly undertaken in 1992-1993 between NRSA, BES, MOA and the Revenue and Land Records departments from the government of Andhra Pradesh, Hyderabad; it was done in three selected sample districts, 15 *mandals,* 45 villages and 2,475 survey numbers. Selection was made based on probability proportional to NAS.

The results of the reconciliation survey indicate that the area under NAS shown in the village records was under-reported to the extent of 10-15 per cent. The survey also shows that the area under "non-agricultural use" derived from remote sensing is 3 to 5 per cent less than the area reported by BES. This is because of the omission of small water bodies and transport networks (small scale of mapping and resolution of satellite data used) and commission of their area under the dominant use class, i.e. agriculture. Hence, 5 per cent of area from agriculture land has been added to the area under non-agricultural use for comparison. The pasture land turned out to be non-existent because it had been converted to agriculture.

F. Wasteland mapping in India

The 1:1,000,000-scale wasteland maps of the states and union territories of India provide a gross estimation of wastelands and their spatial distribution. Because of the small scale and low resolution of the data used (80 m) any isolated patch of less than 100 hectares could not be mapped. Moreover, the small scale also meant that the above maps could not be used for any massive reclamation at the micro-level.

In 1985, the prime minister of India constituted a National Wastelands Development Board (NWDB) with the objective of bringing 5 million hectares of land every year under fuelwood and fodder plantations through a massive programme of afforestation, tree planting and other economic uses. Such development efforts needed very reliable data about the type, extent, location and ownership of wastelands on a large scale.

1. Description of wasteland

Wastelands are degraded lands that can be brought under vegetative cover with reasonable effort, and which are currently under-utilized, and lands that are deteriorating from lack of appropriate water and soil management or on account of natural causes. Wasteland can result from inherent/ imposed disabilities due to such factors as location, environment, chemical and physical properties of the soil or financial or management constraints.

2. Wasteland classification system

Confronted by varying estimates by different agencies, including data thrown up by the latest technical tools, like remote sensing, it became evident that NWDB should lay down precise definitions of the various categories of wasteland. Therefore a technical task force group was constituted by the Planning Commission and NWDB to arrive at precise definitions and categories. The classification system recommend by the technical task force group consists of 13 categories of wasteland:

(a) Gullied and/or ravinous land;
(b) Land with or without scrub;
(c) Waterlogged and marshy land;
(d) Land affected by salinity/alkalinity;
(e) Shifting cultivation area;
(f) Under-utilized/degraded notified forest land;
(g) Degraded pastures/grazing land;
(h) Degraded land under plantation crop;
(i) Sands (coastal and desertic);
(j) Mining/industrial wastelands;
(k) Barren rocky/stony waste/sheet rock area;
(l) Steep sloping area;
(m) Snow-covered and/or glacial area.

3. Wasteland mapping

According to the wasteland statistics generated from the maps prepared by NRSA, 16 per cent of the country's geographical area is under wasteland. Hence, 146 districts having more than 15 per cent of the geographical area under wastelands were considered as critical (except border districts) and selected for mapping in Phases I and II. In Phase III, 88 districts of Madhya Pradesh, having 5-15 per cent of the geographical area, were considered for mapping, and in Phase IV the remaining seven districts were considered.

Since the workload in the project is enormous, all the state remote sensing centres and some central organizations were involved in the mapping task. To maintain a standard and uniform approach to mapping, a manual of wasteland mapping was prepared, giving all the details of mapping and final drawing. Also training cum workshops were organized at NRSA to train all the scientists involved.

For large-scale mapping, both Landsat Thematic Mapper data and Indian Remote Sensing data were used. Visual interpretation of enlarged satellite (1:50,000 scale) false colour composite images based on image characteristic such as colour, texture, pattern, shape, size, location and association were used to identify and delineate different types of wastelands. After preliminary interpretation, field checks were carried out in selected doubtful areas and necessary corrections were incorporated. After the final interpretation, the wasteland thematic details were transferred to base maps prepared from the topographical maps at 1:50,000 scale, using optical instruments.

The village boundaries available in *taluk* maps and forest compartment boundaries from forest compartment maps were transferred on to the wasteland map using optical instruments. Hence, the final outcome is a wasteland map with village boundaries and forest compartment boundaries at 1:50,000 scale. It is possible to identify the villages with wasteland using these maps, and to determine the possible ownership after comparison with cadastral maps available at 1:4,000 and 1:8,000 scale. For the maps being prepared in Phases III and IV, boundaries of watersheds that are 5,000-8,000 ha in area are being incorporated; they are delineated based on AIS and LUS atlas Survey of India toposheets. These maps will help in undertaking development activities based on watersheds.

4. Results

About 64 per cent of the total geographic area, i.e. 203,850,000 ha, has been covered for 1:50,000 scale wasteland mapping in Phases I and IV. According to the wasteland maps prepared in the project, about 35,650,000 ha, i.e. 17.49 per cent of the total geographic area, is lying as wastelands. The final wasteland maps will contain thematic details from satellite data, base map details from topographical maps, village boundaries from *taluk* maps, and reserve forest boundaries from forest compartment maps. About 4,000 wasteland maps of 241 districts have been prepared and being sent to various state users for wasteland reclamation measures.

G. Urban applications using remote sensing techniques

There are a variety of urban applications in which satellite-based, remotely sensed data is being applied, such as (a) base mapping; (b) urban sprawl/urban spatial growth trends; (c) mapping and monitoring urban land uses; (d) urban change detection; (e) urban utility and infrastructure planning; (f) urban land use zoning; (g) urban environment and impact/hazard assessment; (h) urban hydrology; (i) urban management models; (j) census or urban population estimation; (k) urban green belt open space mapping. Below are described the major areas of urban applications in India.

1. Base mapping

In town or city planning the term "base maps" (showing base spatial details of physical and cultural features at derived accuracy and scales) is normally applied to large-scale maps in the range of 1:500 to 1:400 to 10,000. Today, mapping at these scales (master plans and zonal land-use plans) is carried out by aerial photography (orthophotos) or by terrestrial surveying. For base maps and thematic maps at 1:12,500, 1:25,000, 1:50,000 and 1:250,000 scale, satellite imagery and Survey of India topographical maps are used for creating or updating base maps for regional and district planning. Consequently, out of 4,689 urban settlements, only 55 towns have large-scale maps (1:20,000), in the form of guide maps (1991 census). Out of these, a few mega-cities like Hyderabad, Bangalore and Madras have detailed base maps at 1:10,000 scale prepared from large-scale aerial photographs.

2. Urban sprawl/urban spatial growth trend analysis

Urban sprawl and urban growth indicate the overall development pattern of a town or city over a period of time, and the information is essential to carry out trend analysis for planning its future. Synoptic view and repetitive coverage is ideally suited for monitoring urban growth patterns, rate of growth urban development onto agricultural land and water bodies, among other phenomena. A sprawl study carried out by NRSA for Hyderabad, Madras and Nagpur, using multi-sensor and multi-date satellite data, shows that the spatial growth of Hyderabad increased from 245 sq km in 1973 to 522 sq km in 1991 (more than double, with 27.4 per cent built-up area coverage); the annual rate between 1981 and 1991 is 4.95 per cent. During the same period, the agricultural land decreased from 785 sq km in 1973 to 685 sq km in 1991, showing a loss of 100 sq km or 12.8 per cent; also, the surface water tank area decreased from 118 sq km to 112 sq km, showing a loss of 68 sq km or 5.9 per cent. Similarly, the urban growth of Madras increased from 188 sq km in 1971 to 491 sq km in 1989 (with 39.75 per cent built-up area coverage); the annual growth rate was 9.16 per cent, and the population was 5.5 million at 2.45 per cent annual growth rate (1981-1991). The agricultural land decreased from 791 sq km (1974) to 488 sq km (1988), a loss of 38.3 per cent, and the surface water tank area decreased from 47 sq km to 42 sq km, a loss of 5 sq km or 9.7 per cent. Similar work on urban sprawl was completed for Bombay and Ahmedabad by SAC in 1991 and Delhi, Lucknow and Jaipur by the Indian Institute of Remote Sensing (IIRS) in 1992. An urban sprawl study of Allahabad and Port Blair has also been completed by NRSA (1990).

3. Mapping and monitoring urban land uses

Urban land use/land cover is a very important input in urban use planning. To minimize the conflicting demands of land utilization and to help in making important decisions on the use of land arising from increases in population and rapid growth of cities, it becomes essential to have baseline information on changing patterns of land use/land cover. In this regard, NRSA, in discussion with urban development authorities and town planning departments, developed an "Urban Land Use/ Land Cover Classification" in 1986-1989 for use with satellite data, which has been adopted in the DOS Remote Sensing Applications Mission project titled "Mapping and Monitoring Urban Sprawl". The existing urban land-use/land-cover maps of Hyderabad and Madras completed by NRSA at 1:50,000 scale had been successfully utilized in the revision of master plans attempted by the Hyderabad Urban Development Authority and MMDA in 1993 and 1992 respectively. Similar outputs are available for Bombay, Ahmedabad, Calcutta, completed by SAC, and for Delhi, Lucknow and Jaipur by IIRS.

The number and types of different land categories mapped would depend on the particular setting of the town or city and the information requirements. Generally, between 15 to 20 land-cover types can be mapped using satellite images, which include details on residential, industrial, transport, institutional, recreational or vacant lands/layouts, as well as utilities, green belts and other features.

H. Conclusions

Land use/land cover mapping using IRS-1A LISS-I data has been a unique exercise for three major reasons. First, it has served to highlight the special merits of IRS data for land use mapping under Indian conditions. Secondly, it has allowed opportunities to develop operational methodologies, both for visual interpretation and digital analysis. And finally, completion of mapping in 442 districts in the "record time" of one year (effectively operational) has given researchers a sense of accomplishment and given them a high degree of confidence to address similar projects of national importance. Overall, it can be concluded that IRS-1A data has come of age in India, providing a reliable and cost-effective means of carrying out land-use resource inventories.

Mapping of wastelands at 1:1,000,000 scale reveals that remote sensing can be a valuable tool for resource surveys of large areas in considerably short time. Large-scale wasteland maps (1:50,000 scale) prepared with enlarged satellite data for critically affected districts of the country will be a valuable tools to achieve the goals of the National Wasteland Development Board. This information, when used with the information of other natural resources like soil, water, geology and slope will help in the optimum utilization of wastelands either for agriculture, afforestation/plantations or pasture development. Remote sensing techniques can also be used in the periodic monitoring of wasteland reclamation programmes.

The future urban information requirement will be manifold and specific to issues relating to residential and industrial areas, infrastructure and utilities, drainage/sewage, water supply, the environment, tourism and other issues; maps will be made on large scales with greater detail and with higher accuracy. In this regard, satellites like IRS-1C with high spatial resolution and multi-spectral data will be ideal for urban planning and management.

References

NRSA (1989). Manual of nationwide land use/land cover mapping using satellite images, Part I, pp. 1-58.

NRSA (1990). Mapping and monitoring urban sprawl of Madras. Unpublished project report, pp. 1-50.

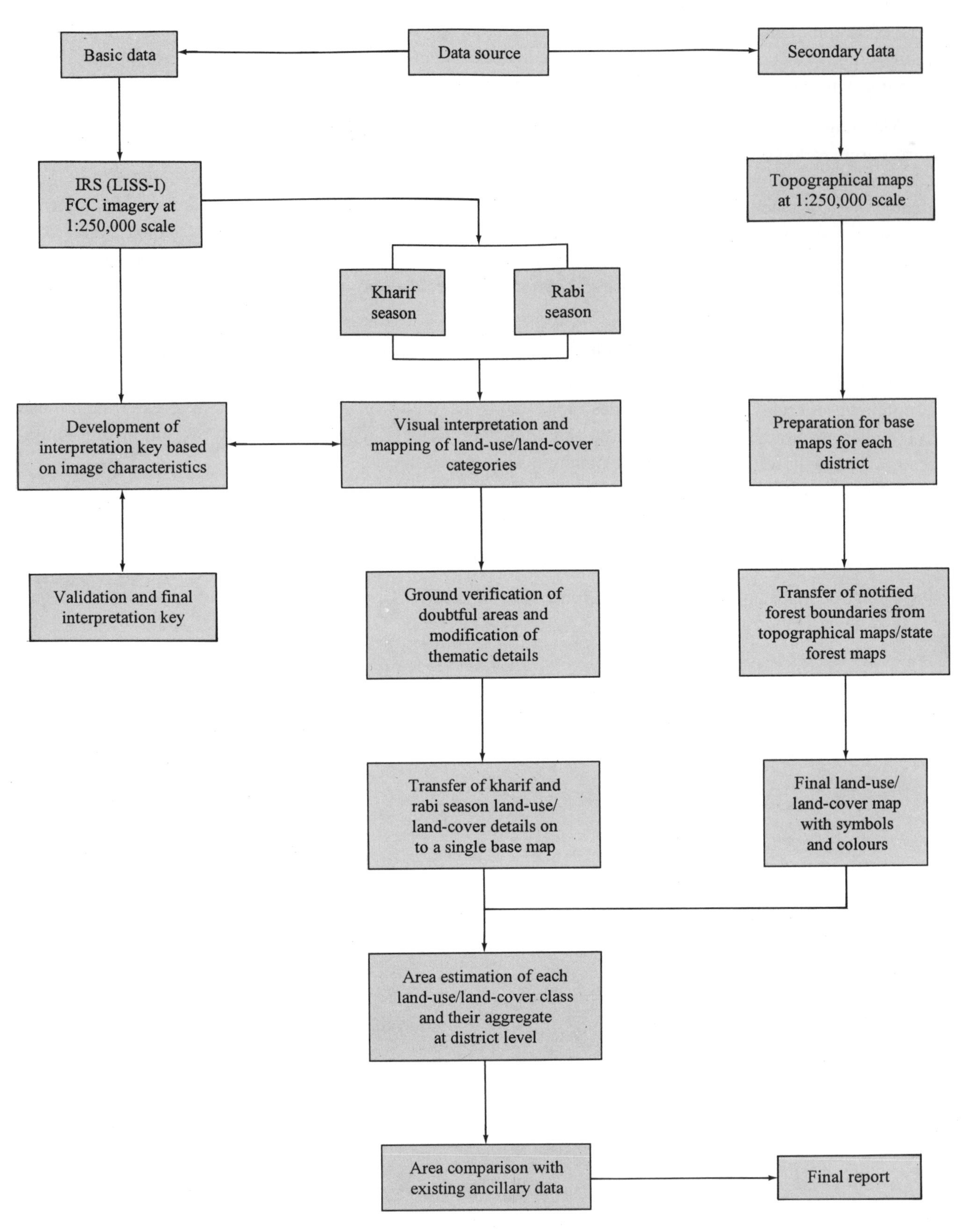

Figure 1. Flow chart showing methodology for mapping land use/land cover through visual interpretation techniques

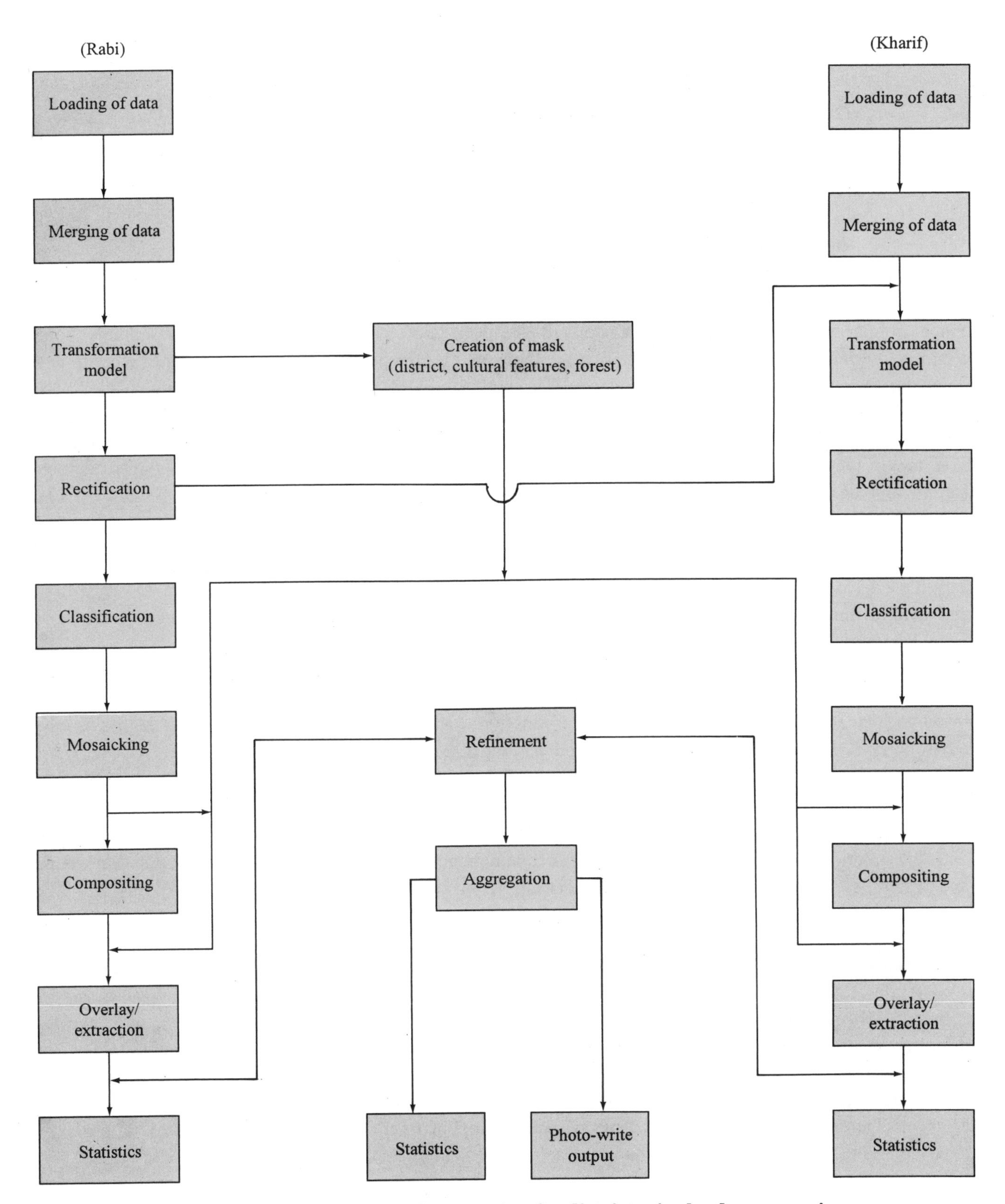

Figure 2. Functional flow chart of digital processing for district-wise land use mapping

APPLICATION OF REMOTE SENSING IN LAND USE MAPPING AND PLANNING

*Tej Pal Singh**

A. Introduction

Land is a primary natural resource which is required for various purposes like agriculture, afforestation/plantation, urbanization and many other developmental activities to satisfy human needs. Since human needs always show an upward trend because of an increase in their numbers and continuous improvement in living standards, their main effects have been to promote growth and development. Human actions have resulted in both use and misuse of the resources available to them. The misuse of land, which may be due to negligence or lack of knowledge and foresight, creates degradation of resources and ultimately affects productivity and the environment adversely. The major land degradation problems faced in India are soil erosion, waterlogging, salinity, deforestation and desertification.

In India, population has been increasing at an alarming rate. This is resulting in competing demands for land for industries, human settlements, irrigation and infrastructure projects, and mining activities, to name only a few. To meet the need of these sectors, developers use up mostly prime agricultural and/or forest lands. Therefore, in such a situation we will have to produce more crops from less agricultural land than we have at present.

It is, therefore, necessary to adopt land use planning on a scientific basis, along with better resource management techniques, so that productivity and incomes improve without degrading the environment. Some of the measures required in this direction are:

(a) Reclamation/development of wastelands;

(b) Proper allocation of land amongst different production and consumption uses;

(c) Periodic evaluation of the dynamic components of land like soil health, vegetation status and water use, and corrective management actions to bring stability to land use;

(d) Regulation of development activities through impact assessment models so that consequences are predicted and only proper actions are executed.

If all of the aspects discussed above are to be incorporated, an integrated approach needs to be adopted. To generate integrated plans for development/resource management of an area, one requires information on functional aspects and interrelationships among different elements of the terrain. For the success of any plan, it is necessary that it should be based on comprehensive, reliable and up-to-date inforrnation. Therefore, land-use surveys are of fundamental importance for land development planning. This paper discusses various components of land use planning and the role of new techniques of remote sensing and geographic information systems (GIS) in the process.

B. Land-use/land-cover surveys

1. Conventional survey

Conventionally, annual information on land use/land cover is collected and compiled at the village level using the nine-fold classification system (table 1). This data is further aggregated at different administrative levels (*tehsil,* district, state and country) and statistics are generated.

* Remote Sensing Applications Group, Space Applications Centre, Indian Space Research Organization, Ahmedabad.

Table 1. Definition of land use classifications

Classification	Definitions
1. Forests	Area under forests includes all lands classed as forests under any legal enactment dealing with forests or administrated as forests, whether State-owned or private, and whether wooded or maintained as potential forest land. The area where crops are raised in the forest and grazing lands or areas open for grazing within the forest are included under the forest area.
2. Area under non-agricultural uses	All lands occupied by buildings, roads and railways or under water, e.g. rivers and canals, and other lands put to uses other than agricultural.
3. Barren and unculturable land	This covers all barren and unculturable land like mountains, deserts, and others; land which cannot be brought under cultivation unless at a high cost shall be classed as unculturable, whether such land is in isolated blocks or within cultivated holdings.
4. Permanent pastures	These cover all grazing lands and other grazing whether they are permanent pastures lands and meadows or not; village common grazing lands are included under this head.
5. Miscellaneous tree-crops and groves not included in the net area sown	Under this class is included all cultivable land which is not included under net area sown, but is put to some agricultural use; lands under casuarina trees, thatching grass, bamboo bushes and other groves for fuel or other uses, which are not included under orchards, are classed under this category.
6. Culturable waste	This includes all lands available for cultivation, whether not taken up for cultivation or taken up for cultivation once, but not cultivated during the current year and last five years or more in succession; such lands may be either fallow or covered with shrubs and jungle, which are not put to any use (they may be assessed or unassessed and may be isolated blocks or within cultivated holdings); land once cultivated but not cultivated for five years in succession is also included in this category at the end of the five years.
7. Fallow land other than current fallows	This refers to all lands which were taken up for cultivation but are temporarily out of cultivation for a period of not less than one year and not more than five years; the reasons for keeping lands fallow may be either poverty of cultivators or inadequate supply of water or malarial climate or silting of canals and rivers or unremunerative nature of farming.
8. Current fallows	This class comprises cropped areas which are kept fallow during the current year; for example, if any seeding area is not cropped again in the same year it is treated as current fallow.
9. Net area sown	This represents the area sown with crops and orchards, counting areas sown more than once in the same year only once.

Source: Adapted from Report of the National Commission on Agriculture, 1976, Part XIV, pp. 149-150.

Classes under the system are based on different criteria such as ownership (forests), use (non-agricultural use), cover (fallow land) and others. Therefore, the scheme fails to provide information about the actual condition of the resource. For example, the category "forests" signifies the area which is legally under the control of the forest department rather than its status. Therefore, this information has limited use in integrated land use planning.

Although various organizations are involved in generating land-use maps, map-making suffers from the problem of limited coverage, age of the data, consistency and so on. Furthermore, most of those maps have been generated to serve the specific objective of the respective organizations.

2. Satellite-based survey

(a) Earth observation systems

Earth observations from space provide data on the natural resources. Large area coverage provides useful insights required to understand complex interrelationships between various natural resources of different regions. Repetitive coverage of the area helps in understanding changing physical processes and helps in closely monitoring the changes taking place in different areas. The impact of development schemes on restoration of vegetation cover of the Earth can also be monitored. Availability of data at different spatial resolutions (as coarse as one km or as fine as 20 m in multispectral mode/5.8 m panchromatic) provides a means for observing the Earth both at a macro level and in detail.

The capability of sensor systems, with respect to providing details of land features, is continuously improving. In the beginning of satellite remote sensing (Landsat-1, in 1972) one hectare on the ground was covered by six data points (80 x 80 m area per pixel). By 1986, thc pixel size achieved in the SPOT satellite was 20 x 20 m (multispectral) and 10 x 10 m (panchromatic), which provided 25 and 100 data points per ha, respectively. With the launch of IRS-1C (1995), images in the panchromatic band provide about 290 data points/ha. It is expected that in the near future these capabilities are going to improve even further (about 10,000 data points/ha).

(b) Classification system

A conventional classification scheme is incompatible with remotely sensed data because remotely sensed data provide information on land cover status that, through logical interpretation, is translated to land-use classes. Therefore, a need was felt to evolve a suitable land-use/cover classification system that could be compatible with both conventional data collection as well as remote sensing techniques.

However, images provided by the different types of space-borne sensors at varying heights do not provide the same information about the land features. Therefore, it is believed that no single classification could be used with all types of imagery (Nunnally, 1974). A general purpose, resource-oriented classification scheme compatible with remote sensing data has been developed by Anderson et al. (1976) (table 2). Under the scheme, a flexible system providing different levels of classification is suggested. Anderson's classification scheme has been modified in many cases to suit local conditions and objectives of different studies. In India, the classification system used for the nationwide land-use/cover mapping project was developed by the Department of Space (National Remote Sensing Agency, Hyderabad) in consultation with user agencies (Anonymous, 1990a) (table 3).

(c) Mapping techniques

Remotely sensed data is being used operationally for land use/cover mapping and monitoring. The basic strength of remotely sensed data is spatial comprehensiveness. Data is available in the form of images which after interpretation are translated into maps. Both digital and visual interpretation techniques are employed to generate the maps. A brief description of the procedures employed in nationwide operational project in India is given below (Anonymous, 1990a, 1990b).

(i) Digital techniques

A schematic flow chart of the methodology is given in figure 1. The procedure comprises the following steps:

(1) *Image restoration:* At this stage the image is corrected for distortions. The image is co-registered with a standard map (Survey of India topographical maps) through a map-image transformation model using ground control points;

(2) *Stratification:* The study area is divided into forest and non-forest strata to avoid the misclassification between forested land and agricultural crops. For this purpose forest boundaries are taken from the available maps;

Table 2. Land-use and land-cover classification system for use with remote sensor data

Level I	Level II
1. Urban or built-up land	11 Residential 12 Commercial and services 13 Industrial 14 Transportation, communications and utilities 15 Industrial and commercial complexes 16 Mixed urban or built-up land 17 Other urban or built-up land
2. Agricultural land	21 Crop land and pasture 22 Orchards, groves, vineyards, nurseries and ornamental horticultural areas 23 Confined feeding operations 24 Other agricultural land
3. Range land	31 Herbaceous range land 32 Shrub and brush range land 33 Mixed range land
4. Forest land	41 Deciduous forest land 42 Evergreen forest land 43 Mixed forest land
5. Water	51 Streams and canals 52 Lakes 53 Reservoirs 54 Bays and estuaries
6. Wetland	61 Forested wetland 62 Non-forested wetland
7. Barren land	71 Dry salt flats 72 Beaches 73 Sandy areas other than beaches 74 Bare exposed rock 75 Strip mines, quarries and gravel pits 76 Transitional areas 77 Mixed barren land
8. Tundra	81 Shrub and brush tundra 82 Herbaceous tundra 83 Bare ground tundra 84 Wet tundra 85 Mixed tundra
9. Perennial snow or ice	91 Perennial snow fields 92 Glaciers

Source: Reproduced from Anderson et al., 1976.

(3) *Classification:* Separate classification operations are performed for two strata. Supervised classification is employed for the non-forest strata using maximum likelihood classifier. The normalized vegetation index {NDVI = (NIR – Red)/(NIR + Red)} is used for the strata corresponding to the forest class. Since NDVI is directly related to vegetative vigor, it provides information about the density of the forest. Thresholds of NDVI for the various forest classes are decided on the basis of area-specific field data/observations;

(4) *Refinement:* The logic is applied to modify discrepant classification resulting from spectral similarity of two or more classes in a non-forest stratum. For this purpose, data from two seasons (kharif and rabi) are classified separately, and the assigned classes of the corresponding pixels are compared. Classification of a particular category in one season is assumed to be correct and

Table 3. Land-use/land-cover classification system

Level I	Level II
1. Built-up land	1.1 Built-up land
2. Agricultural land	2.1 Crop land (i) Kharif (ii) Rabi (iii) Kharif + rabi 2.2 Fallow 2.3 Plantation
3. Forest	3.1 Evergreen/semi-evergreen forest 3.2 Deciduous forest 3.3 Degraded or scrub land 3.4 Forest blank 3.5 Forest plantation 3.6 Mangrove
4. Wasteland	4.1 Salt-affected land 4.2 Waterlogged land 4.3 Marshy/swampy land 4.4 Gullied/ravined land 4.5 Land with or without scrub 4.6 Sandy area (coastal and desertic) 4.7 Barren rocky/stony waste/sheet rock area
5. Water bodies	5.1 River/stream 5.2 Lake/reservoir/tank/canal
6. Others	6.1 Shifting cultivation 6.2 Grassland/grazing land 6.3 Snow-covered/glacial area

Source: Anonymous, 1990a.

based on observation and logic, so the pixels are assigned new classes in data for the other season. For example, a pixel classified as "plantation" in rabi and "crop" in kharif is refined as plantation in kharif. Similarly, "plantation" in kharif and "fallow land" in rabi necessitates change in the kharif pixel class to "crop";

(5) *Overlaying of cultural features:* Various cultural features (roads, rails, canals and so on) and administrative boundaries are digitized from the base maps, which are overlaid on to the classified output;

(6) *Statistics generation:* Administrative unit-wise aggregation of each class is done, and statistics for land-use/cover classes are generated.

(ii) Visual techniques

Mapping of land use and land cover from satellite imagery through visual interpretation techniques is well established. A procedure for mapping land use/land cover through visual interpretation techniques is described in a schematic diagram (figure 2):

(1) *Selection of satellite data:* The season and the product/format of satellite data is selected on the basis of scale of mapping, level of details required, unit of mapping, availability of equipment and other considerations;

(2) *Ground truth/collateral material:* After the images of interest are studied, a field visit is performed to collect information about various land-cover classes and their conditions. Other data in different forms (map, statistics, graphs and so on) from many sources, like literature, measurements and analysis, and ground and aerial photographs, are used to aid and verify interpretation;

(3) *Interpretation keys:* Area-specific, class-wise interpretation keys are developed on the basis of image characteristics and ground truth. After they are applied to selected areas, the keys are tested and modified accordingly;

(4) *Base map preparation:* Information related to land use/land cover derived from remotely sensed data provides polygons belonging to different categories. However, it is always necessary to find the locations of these units with respect to administrative boundaries and transport (road, railway or water) links. This is achieved by taking the details about habitation, road/railways and water bodies from Survey of India maps at 1:250,000 or 1:50,000 scale. A map providing these details is called a base map;

(5) *Interpretation:* Some permanent features from the base map are matched with the same features on the image. Then the image is classified into different categories based on interpretation techniques/keys described earlier. Doubtful areas/categories are identified and other data sources are studied to overcome confusion. A persistently doubtful area is again visited and interpretation is verified. Modifications are made accordingly and the final map is prepared;

(6) *Accuracy estimation:* Errors in a map prepared through interpretation of remotely sensed data may result from (a) inherent inaccuracy in source data (locational accuracy), (b) limitations of cartography and printing (mapping accuracy) or (c) wrong interpretation (interpretation or classification accuracy). Users are most concerned with classification accuracy. In general, statistical methods are used to evaluate classification accuracy. The procedure involves selection of samples of the required size, checking in the field, generation of error matrix (also called confusion matrix) and estimation of class-wise and overall accuracy;

(7) *Area estimation:* Acreage estimation under different classes is done from the maps, and statistics are generated according to administrative (district) or natural (watershed) boundaries.

C. Nationwide land-use/land-cover mapping project

The nationwide land-use/land-cover mapping project was taken up under the aegis of the Department of Space and at the instance of the Planning Commission of the Government of India. The main objective of the project was to prepare land-use/land-cover maps at 1:250,000 scale for generation of operational plans for all 15 agro-climatic zones of the country. It was also envisaged to provide district-wise agricultural statistics for kharif and rabi seasons. The project was coordinated by the National Remote Sensing Agency, and 26 organizations from central and state governments participated.

Two-season (kharif and rabi) data (73 m spatial resolution) from an Indian Remote Sensing (IRS) satellite was used for mapping; both digital and visual techniques were employed. The districts with relatively plain terrain (168 in number) were mapped using digital techniques, and 274 districts through visual technique. The maps were prepared at a scale of 1:250,000. The results of the land use/cover mapping take the form of statistics, line maps/colour-coded photo-write outputs, and reports. The land use statistics for the country are provided in table 4, and a typical land-use/land-cover map is shown in figure 3.

D. Land use planning: an integrated approach

The definition of land given by the Food and Agriculture Organization (1976) implies that it should be studied as an operational subject, in preference to separate analytical investigations of its component parts. Justification for an integrated approach is threefold (Townshend, 1981):

(a) Available knowledge from different sciences suggests that functional interrelationships between various components of land, such as hydrology, pedology, geomorphology and land use, are very close;

(b) Particular characteristics and processes at a location usually extend to some distance, hence providing the possibility of regionalization;

Table 4. Area under detailed land-use/land-cover categories, India (1988-1989)

Category	Area in ha	Percentage of total geographic area
Built-up land	13,913,772	04.24
Agricultural land		
Kharif land	120,586,769	
Rabi crop land	76,300,445	
Double-cropped area	53,105,792	
Net area sown ($)	151,481,769	46.09
Fallow land	13,762,590	04.17
Forest		
Evergreen/semi-evergreen	14,184,366	04.32
Deciduous forest	31,813,875	09.68
Degraded forest	16,274,270	04.95
Forest blank	1,813,853	00.55
Forest plantations	1,119,452	00.34
Mangroves	504,999	00.15
Wasteland		
Salt-affected	1,988,380	00.60
Waterlogged	1,219,666	00.37
Marshy/swampy	823,876	00.25
Gullied/ravined	2,020,329	00.62
Land with or without scrub	26,514,564	08.06
Sandy area (coastal and desertic)	5,572,086	01.69
Barren rocky/stony waste/sheet rock area	6,251,414	01.91
Water bodies		
River/stream	8,414,852	02.55
Lake/reservoir/tank/canal	2,195,968	00.67
Others		
Grassland/grazing land	3,104,538	00.94
Mining/industrial waste	116,497	00.03
Shifting cultivation	2,823,626	00.87
Snow-covered/glacial area	6,992,255	02.14
Salt pans	37,808	00.01
Unclassified	7,452,095	02.27
Unsurveyed (*)	8,329,400	02.53
Total	328,726,300	100.00

Source: Report on area statistics of land use/land cover generated using remote sensing techniques, 1995, NRSA, Hyderabad.

Note: ($) Includes 7,700,343 ha or 02.34 per cent of area also under agricultural plantations.
(*) Unsurveyed area is in Jammu and Kashmir due to hanging boundaries.

(c) In keeping with the justifications above, one may use a limited number of observations to estimate many other characteristics and qualities.

Remotely sensed data provides information in spatial format; therefore, it is best suited for extrapolation from a limited number of observations to the whole area.

E. Planning model

It is recognized that land use should be investigated from all viewpoints related to cultural, socio-economic and ecological conditions. Land-use planning models, therefore, make use of such data. A number of components, from data acquisition to decision-making, have been enumerated for this purpose (Joyce, 1973).

Steinitz (1993) suggested six levels of enquiry (models) in landscape planning:

(a) *Representation:* At the first level of enquiry, the various elements/parameters characterizing the landscape are used to describe land (use) in context, space and time. The data acquisition for this purpose may done from different sources;

(b) *Process:* In the second level of the enquiry, the functional and structural relationships among different terrain elements are studied. Data/information available from the first model is used for the purpose;

(c) *Evaluation:* At this level of enquiry, the assessment of functions is done. This helps answer questions such as, "Is the current landscape functioning well in terms of health, cost and user satisfaction?" Different criteria and models (run-off and soil loss, for example) are used as analytical tools;

(d) *Change:* Depending upon the evaluation results, one may need to alter/change the process through alternate strategies. The strategy may be suggested on the basis of the information available from the representation model;

(e) *Impact:* A possible impact assessment of suggested strategies must be done before actions are initiated. A simulation model may be used for this purpose;

(f) *Decision:* Based on all these enquiries a decision is taken and plans are formulated.

To implement these models in a real life situation, there is a need for various data sets related to different themes. Remote sensing and GIS can play important roles in providing information in spatial contexts for various resources (themes) and in their analysis, respectively.

F. Information level vs application

Various levels of detail (and at different scales) generated through remotely sensed data have a direct relevance to the scale of planning. In general, level I/II (about 1:1,000,000/1:250,000 scale) information is used for national- and state-level policy formulation and perspective planning; level II-type information (1:50,000 scale) is suitable for project formulation; level III/IV information (about 1:15,000 scale) is required for micro-level planning and for generating implementation plans. Today, satellite data is available from different sensor systems designed for specific applications. Users may select the satellite data products according to their requirements and objectives.

The scale of mapping and level of details that can be derived from satellite images depend on the characteristics of the sensor system. The multiple sensor system on board IRS-1C is providing data in different spectral bands at various spatial resolutions. Land use/land cover mapping can be done at various scales and levels using different sensor data from the same space platform. Wide-field sensor (WiFS) data can provide level I information for regional level applications, whereas LISS-III in combination with PAN can be used for cadastral level information (Rao et al., 1996).

G. Case studies

A large number of studies related to integrated land use planning have been conducted by various organizations in different part of the country using remotely sensed data. Here two studies, one at macro-level (Panchmahals) and the other at micro-level (Dhandhuka), are briefly described.

1. Panchmahals

Panchmahals is one of the economically and industrially backward districts of Gujarat and is dominated by a tribal population. Although the average rainfall of the district is about 900 mm, many regions face shortage of water because of high run-off due to undulating terrain, shallow soils and rapid deforestation.

A case study was carried out to support decision makers in planning related to natural resource management at the district level (Anonymous, 1991; Ghosh et al., 1993). The study was carried out at 1:250,000 scale. The basic data source used for the analysis of the natural resources was false colour composite from IRS LISS-I (73 m resolution).

Thematic maps on land use, groundwater prospect and soil were prepared using these space images. Slope information was derived using topographical maps. These maps were integrated using geographic information system techniques, and a map showing composite land development units (CLDU) (homogeneous in physical characteristics) was prepared for the district. After obtaining various composite land development units, each characterized by a set of attributes with respect to slope, land use, soil and groundwater prospect, watershed boundaries were superposed, and sediment yield index for each CLDU was derived. Depending upon the nature of CLDU and silt yield indices, soil conservation plans have been suggested and alternate land use plans have been indicated. Wastelands, which are on relatively gentle slopes and have better soil conditions, have been suggested for agriculture/plantation/grassland development. Those CLDUs that are wastelands within the notified forest area have been suggested for afforestation. The study provided *taluka*-wise plans for soil and water conservation, and alternate land uses for the whole district.

2. Dhandhuka

A detailed study for a micro-watershed (3,000 ha) was carried out in the *taluka* of Dhandhuka in Ahmedabad District, Gujarat (Singh et al., 1996). The study area is characterized by a semiarid climate. The average annual rainfall for the last 10 years is 495 mm. The area is drained by small seasonal nalas. The agriculture is mainly rain-fed, with predominantly cotton crop occupying fields in both kharif and rabi seasons. The area is occupied by fluviomarine deposits with a thin veneer of soil of sub-recent to recent age experiencing the problems of soil erosion and soil salinity. Geomorphologically, it is part of a palaeo-deltaic plain, which is indicative of reasonably good groundwater potential but brackish to saline quality.

As the first step towards land resource development planning, land transformations in the watershed over the last 100 years were studied. Survey of India topographical maps at 1:63,360 scale surveyed during 1968-1974 were taken as base. These maps provide information about the extent of agricultural land and wastelands. The information available on these maps was compared with recent satellite images. It was revealed that a part of the wasteland has been converted to agriculture, and the remaining part continues to be wasteland. During this period some portion of the agricultural land also turned into wasteland. An increase in the drainage density of the watershed during this period indicates the severity of soil erosion.

Poor crop growth in the upper portion of the watershed (as seen in the satellite image) is a result of a lack of sufficient moisture, whereas in the lower portion, land degradation due to floods in seasonal *nalas* is the main cause for poor crop condition. Thematic maps were prepared at 1:50,000 scale using three-season, IRS LISS-II data. Integrated analysis (as discussed in the section above on Panchmahals) of different map layers was carried out. This analysis indicated the need for the following:

(a) Soil and moisture conservation;
(b) Reclamation of wastelands;
(c) Increase in surface water storage capacity;
(d) Proper cropping pattern (agro-horticulture, agroforestry).

Based on site-specific characteristics, the developmental plan for agricultural/land resources has been prepared. An action plan on a cadastral map for a portion of the watershed is given in figure 4.

The plan was verified in the field and discussed with the beneficiaries of the watershed by a non-government organization (BAIF Development Research Foundation, Vadodara). BAIF is now implementing these actions in the field. The District Rural Development Agency, Ahmedabad, has provided funds for this purpose.

H. Conclusions

Land use planning on a scientific basis is a must, to arrest the degradation of the natural resources and improve productivity. Remotely sensed data provide reliable and up-to-date information at different levels in a spatial format; therefore, it is best suited for resource development planning at both the macro-level as well as micro-level. GIS techniques are found to be of great help in analysis of spatial data sets. The plans generated through interpretation of satellite data are being accepted by the users.

References

Anderson, J.R., E.E. Hardy, J.T. Roach and R.E. Witmer (1976). A land-use and land-cover classification system for use with remote sensor data. Geological Survey Professional Paper 964, United States Geological Survey, Reston, Virginia.

Anonymous (1990a). Manual of nationwide land use/land cover mapping using satellite imagery, part I. National Remote Sensing Agency, Hyderabad.

Anonymous (1990b). Manual of nationwide land use/land cover mapping using digital techniques, part II. Regional Remote Sensing Service Centre, Nagpur.

Anonymous (1991). District-level planning: a case study in Panchmahals District, Gujarat. An interim technical report, Space Applications Centre, Ahmedabad, SAC/RSA/NRIS-DLP/TR-6.

FAO (1976). A framework for land evaluation. *Soils Bulletin,* No. 22.

Ghosh, Ranendu, R.K. Goel, B.S. Lole, T.P. Singh, K.L.N. Sastry, J.G. Patel, Y.V. Vanikar, P.S. Thakker and R.R. Navalgund (1993). District-level planning: a case study for the Panchmahals District using remote sensing and GIS techniques. *International Journal of Remote Sensing*, 14: 3163-3168.

Joyce, A.T. (1973). Land use and mapping. In *Proceedings* of the Third ERTS Symposium, Goddard Space Flight Centre, Washington, D.C., 10-14 December, vol. III.

Nunnally, N.R. (1974). Interpreting land use from remote sensor imagery. In J.E. Estes and L.W. Senger, eds., *Remote Sensing Techniques for Environmental Analysis*. Santa Barbara, California: Hamilton.

Rao, D.P., N.C. Gautam, R. Nagaraja and P. Ram Mohan (1996). IRS-1C applications in land use mapping and planning. *Current Science*, 70: 575-581.

Singh, T.P., A.K. Sharma, Tara Sharma, R.R. Navalgund, P.H. Vaidya, P.S. Dwivedi and U.D. Datir (1996). Applications of remote sensing and GIS in integrated watershed development planning: a few case studies in Ahmedabad District, Gujarat. In *Proceedings* of the National Workshop on Application of Remote Sensing and GIS Techniques to Integrated Rural Development, Hyderabad, June 14-15.

Townshend, J.R.G., ed. (1981). *Terrain Analysis and Remote Sensing*. London: George Allen and Unwin.

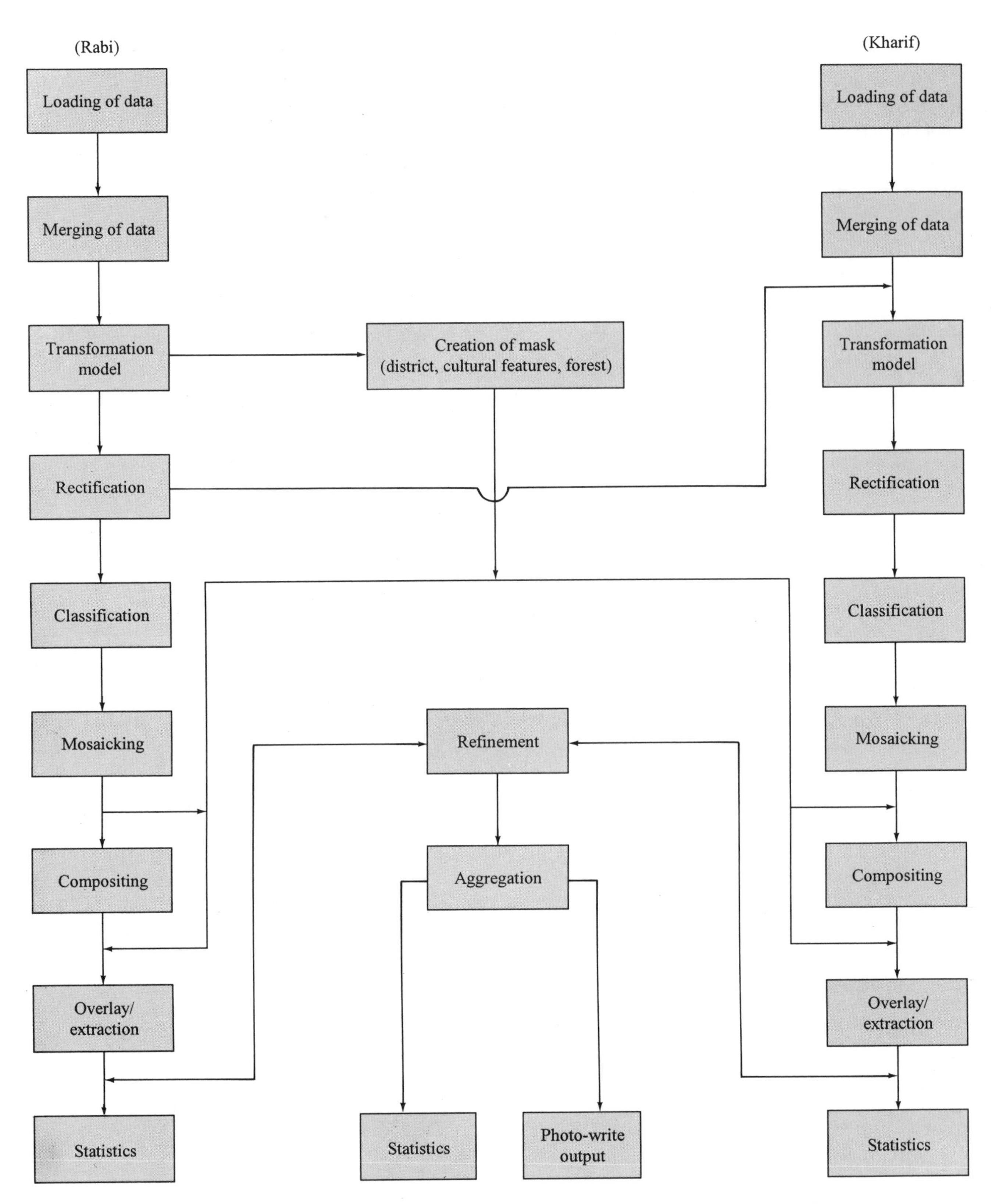

Figure 1. Functional flow chart of digital processing for district-wise land use mapping
(*Source:* Anonymous, 1990b)

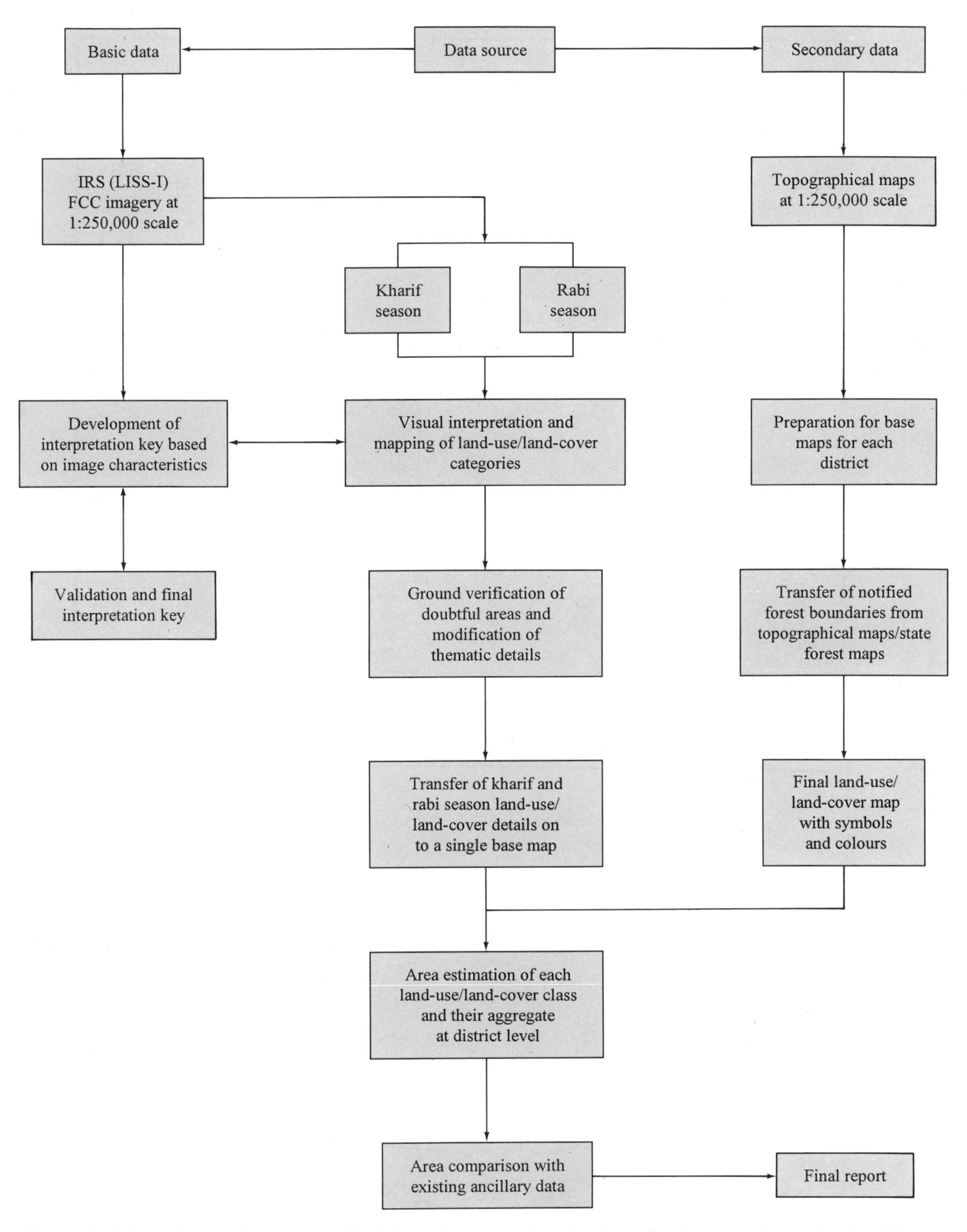

Figure 2. Flow chart showing methodology for mapping land use/land cover through visual interpretation techniques (*Source:* Anonymous, 1990a)

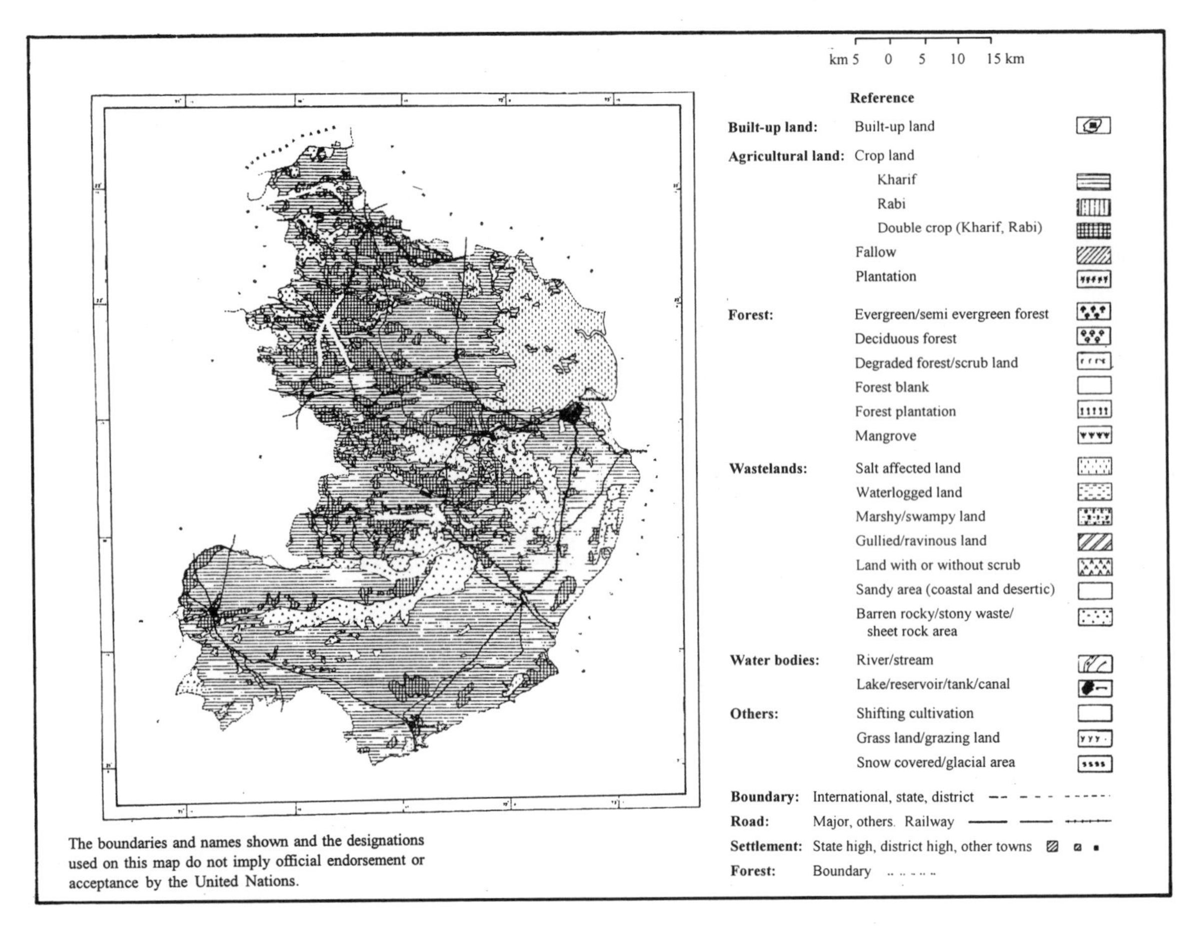

Figure 3. Land-use/land-cover map of Bhavnagar District, Gujarat

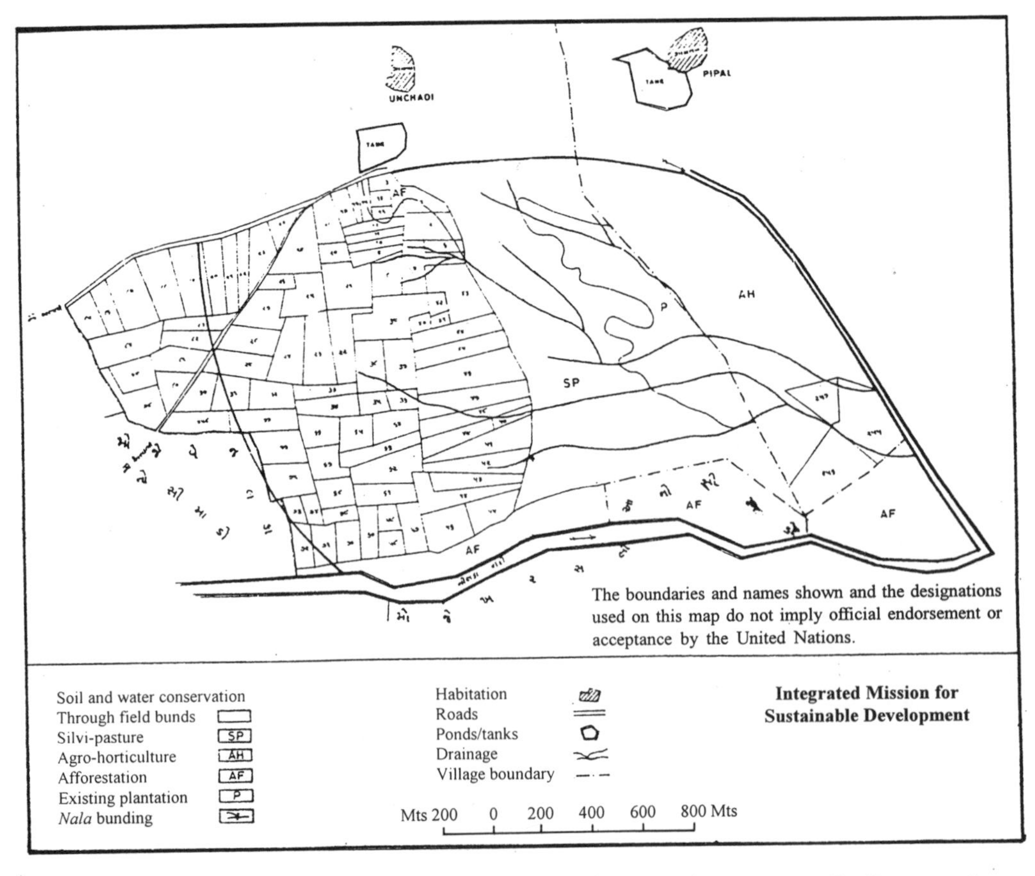

Figure 4. Action plan for land resource development for a micro-watershed on cadastral map in Dhandhuka *taluka,* Ahmedabad

REMOTE SENSING AND GIS APPLICATIONS FOR LAND USE PLANNING IN MYANMAR

*Aung Kyaw Myint**

A. Introduction

GIS technology was introduced to the Myanmar Forest Department with the National Forest Inventory and Management project (MYA/85/003). It was a continuation of the National Forest Inventory project (BUR/79/011). A PC-based Arc Info system was successfully installed and the staff of the Forest Department were trained in the use of GIS facilities; the expertise of the computer staff was later applied for training staff of various departments. The GIS unit of the Forest Department is the first of its kind in Myanmar.

B. GIS facilities available in Myanmar

PC Arc Info version 3.4.1 was installed on three Compaq 80486 computers in July 1993. A Calcomp 9500 A0-size digitizer, a Calcomp 2030 A0-size drum pen plotter and an HP PaintJet plotter were provided.

Since Myanmar is a member of International Centre for Integrated Mountain Development (ICIMOD) some GIS assistance was received. GIS equipment was given to the Forest Department, Department of Border Area and Mountain Area Development, National Commission for Environmental Affairs (NCEA) under the Ministry of Foreign Affairs (MOFA), and Department of National Housing. The Forest Department helped in training the staff of the receiving organizations. ICIMOD strengthened the Forest Department GIS unit with two computer units: an AST Bravo LC 4/66D PC and a Calcomp Drawing Board III digitizer. GEOLINK Mapping software and IDRISSI for Windows software were also given.

UNDP/FAO community development projects were also interested in GIS applications for land use planning in the project areas, and decided to provide some more facilities to the Forest Department's GIS unit. Four Pentium 90 MHz computers, together with one 80486 laptop notebook, an HP DesignJet 650C A0 plotter, HP DeskJet 1200C A4 printer, HP LaserJet 5P, Exabyte 8205 8-mm tape drive, Bausch and Lomb zoom stereo-transferoscope, Sharp A3 colour scanner, and three GARMIN SRVY II portable geopositioning systems (GPS) were added. PC Arc Info version 3.4.2, PCTIN, ARCVIEW version 2.1a, EASI/PACE version 6.0, and dBASE V were included.

To sum up, the following hardware and software were available at the Forest Department's GIS unit, in the Computer Section:

Hardware

Compaq PC 486 66 Mhz, 8 MB RAM, 320 HD	3 units
Calcomp 9500 A0-size digitizer	
Calcomp Drawing Board III digitizer, A1 size	
Calcomp 2030 pen plotter, A0 size	
Pentium 90 Mhz, 32 MB RAM, 1 GB HD	4
Altek A0-size digitizer	
HP DesignJet 650C A0-size plotter	
HP DeskJet 1200C A4-size printer	
HP PaintJet A4-size plotter	
HP LaserJet 5P	
Exabyte 8205 8-mm tape drive	

* Forestry Department, Institute of Forestry, Yezin, Myanmar.

Bausch and Lomb zoom stereo-transferoscope	
Sharp A3-size high-resolution colour scanner	
GARMIN SRVY II portable GPS	3

Software

PC Arc Info version 3.4.2.	5 units
PC TIN	
ARCVIEW version 2.1a	2
EASI/PACE version 6.0	2
dBASE V	

C. GIS applications for UNDP/FAO-assisted community development projects in Myanmar

Lansdat satellite images of February 1995 for some project areas were purchased, including images on magnetic media and printed hard copies.

1. Watershed management for three critical areas (MYA/93/005)

An integrated approach was practiced for this project, in which Landsat images were used for identification of present land use, and aerial photographs were used for detailed mapping and identification. The village tract level was classified for poverty mapping using socio-economic data and land-use maps.

2. Community development of Ayeyarwady mangroves (MYA/93/026)

All the costs of purchasing hardware, software, images and subcontracts for mapping, including the costs of training and material supplies, were shared by two projects interested in GIS applications. The mangrove project mapped the forest cover and land-use changes over 20 years, i.e. between 1974 and 1994. Poverty mapping was not tried because a sound socio-economic database was not available.

D. Other applications

1. Development Planning in Kabaw Valley

A land-use planning exercise for development planning and decision-making in the Kabaw Valley region had been done with the aid of remote sensing data. Manual interpretation of the existing land-use patterns was done and the maps were printed.

2. Land use planning in Bago Yoma region

A land-use planning exercise for the Bago Yoma area, an area famous as the home of teak, was also developed with the assistance of some organizations from Japan. The work is still in progress.

3. Other applications

The following applications were developed with the GIS facilities available at the Forest Department:

(a) Nine Arid District Greening project
Coverage (6): Project, district, township, road, waterbody, town;

(b) Yezin University training forest geographic database
Coverage (6): Township, land-use, forest, village, stream, contour;

(c) Topographic map of Myanmar geographic database
Coverage (4): Water body, contour, town, route;

(d) Poppa geographic database
Coverage (5): Land use, river and stream, forest, route, project area.

E. Future applications and conclusion

The following list shows the provisional plan for future applications:

(a) National-level geographic database;
(b) Sittaung Valley geographic database;
(c) Bago Yoma geographic database;
(d) Dry zone geographic database;
(e) Installation of a GIS training centre at the Institute of Forestry with ICIMOD assistance.

It may be concluded that the GIS unit is well established in the Forest Department and that successful applications are very productive for Myanmar. Computer Section staff are also very competent for the future development of planning through GIS.

IMPROVING A LAND-USE MAP USING GEOGRAPHIC INFORMATION SYSTEMS AND REMOTE SENSING

*Suthep Chutiratanaphan, Bunruk Pattanakanok**,*
*Kees Bronsveld and Harry Kostwinder***

ABSTRACT

This study was done in the context of a combined project of the Land Development Department of Thailand and ITC, Netherlands, with the objective of studying to what extent a GIS could be used for land-use planning. The study area consists of a few sub-districts of the province of Petchabun, northern Thailand. This paper reports on only a part of the study with the objective of producing a detailed, up-to-date land-use map as a basis for land use planning. A soil map and multitemporal satellite images were available. A knowledge base was built up by extensive fieldwork. The knowledge base contained information on the phenology of crops and on the type of land on which different agricultural crops are grown in the study area. Artificial intelligence (AI) techniques were used in a GIS environment to extract information on the present land use. A land-use map was produced showing the distribution of the main crops in the area. The resulting map was compared with the existing land-use map made with outdated aerial photographs.

The resulting map did not have very large discrepancies with the existing land-use map, but did give some advantages. In particular this map showed improvements in the areas where second crops were being cultivated, compared to the land-use map produced with outdated aerial photographs. Also, the localization of areas where maize was cultivated was much more detailed than in the existing map. This method will probably be an improvement on existing methods to process satellite images for land use mapping, because the amount of field work can be reduced and there is improved flexibility to combine other sources of information with reflection data from satellites.

A. Introduction

This study was carried out by the Land Development Department, Thailand, and ITC, Netherlands, as part of a joint research programme. The main objective of this research programme was the development and testing of an operational information system for natural resource management and land use planning for land resource and land-use conditions and data availability situations as they prevail in Thailand.

The main aim of the case study was to investigate the potential of the information system for:

(a) The assessment of sustainability problems (erosion, land degradation) of the current land use;

(b) The (semi-) quantitative assessment of ecological and economic consequences of the introduction of more sustainable land-use alternatives.

In order to do so, information is required, not only on the amount of different crops, but also where they are grown and on what type of land. Therefore an important input in the sutdy was a map showing the rapidly changing land use in the study area. This paper reports on the production

*Land Development Department, Ministry of Agriculture and Cooperatives, Bangkok, Thailand.

** International Institute for Aerospace Survey and Earth Sciences (ITC), Enschede, Netherlands.

of a land-use map of the area, using satellite imagery and a soil map, making use of remote sensing and GIS techniques. For a description of the information system see Huizing and Bronsveld, 1991.

1. The study area

The study area is a part of Lom Kao District in the upper Pa Sak catchment, Phetchabun Province, Thailand. The area consists of a broad valley bordered by sloping and steep hilly and mountainous land. Most of the sloping and steep land has been deforested in the last 20 years and is now used for annual crop production (predominantly maize as a single crop; partly maize followed by mung beans as a second crop). The maize is grown as a cash crop. Downslope ploughing is the dominant land preparation practice.

2. Materials used

- 1:50,000 soil map (LDD, 1990);
- 1:100,000 land-use map (LDD, 1986);
- Landsat TM images of 1988 and 1989;
- Land use and farming systems data (Anaman, 1990);
- Land-use map produced with aerial photographs of 1974 (LDD, 1986).

3. Tools

The processing of data was done on microcomputers with the following software:

- Automated Land Evaluation System (ALES);
- ILWIS/GIS software for analysis of satellite images, overlaying of maps, creation of digital terrain model and preparation of output maps.

4. Disadvantages of current methods

In various reports concerning the mapping of land use using remote sensing, it has been shown that the accuracy of the results can be improved by using multiple satellite images (e.g. Bronsveld and Luderius, 1982). However, with the increasing amount of data, using more images, the effort of combining the data into useful information increased to a level that makes the interpretation of the original satellite images difficult, especially if the agriculture pattern is complicated. To solve this problem, various authors have made a maximum likelihood classification of multiple satellite images (Gangkofner et al., 1990).

This method requires extensive fieldwork, usually conducted with an area frame sampling approach. Not only does much work have to be done in the field, but also the geo-referencing of the collected data to be able to compare them with the satellite data is time consuming. With all crops in the area, all variations in reflection should be sampled; these variations can be caused by different types of soils, by variations in the planting time caused by climate differences within the study area, or by differences in management practices. To compensate for some of these variations the area is stratified in regions where the soils and the climate are assumed to be homogeneous. However, if in the same area an inventory is needed in another year, the same amount of field work must be done, because of differences in recording dates or atmospheric conditions.

Another disadvantage is the difficulty of integrating the knowledge on reflection of crops throughout the year and information from thematic maps (e.g. on soils or climate) with reflection data acquired from satellite images.

B. Methodology

Considerable knowledge is available on the reflection of vegetation, the type and amount of plant cover during the year of different crops and what crops are grown in different soils. Evidence gathered from reflection data of different satellite images and from soil map was available. AI methods have been proven to combine knowledge data into useful information.

1. Dempster-Shafer theory of evidence

The method used in this paper is based on the Dempster-Shafer theory of evidence (see Shafer and Logan, 1987) to combine data from multiple sources, but is considerably simplified for efficient use for large data sets on a microcomputer. In this section the theory is briefly explained, and the next section will explain the method that was followed.

The basic idea of the theory is the breakdown of the large body of data (from multiple sources) into components. For each of these components, probability judgements are made, which are then combined (see Srinivasan and Richards, 1990, for a demonstration of the theory for remote sensing applications).

The following definitions and rules are given here to show the main aspects of the theory:

(a) The set of all possible labels q = {a, b, c} is called the frame of discernment. In our case, the frame of discernment includes all possible labels that are available for classification of the pixels;

(b) Belief that a pixel has a particular label is expressed as a basic probability assignment (bpa), which is a number between 0 and 1. The bpa is often called the "mass" committed to the set, usually denoted by m;

(c) The belief in a set (denoted by Bel) is the sum of each mass committed to its subsets; for example, Bel (q) = 0.9, if m(a) = 0.3, m(b) = 0.4 and m(c) = 0.2;

(d) The empty set can have no mass committed to it ($m(\theta) = 0$);

(e) Also belief against A can be specified, denoted by Bel (–A);

(f) In the theory of evidence, belief is non-additive: Bel (A) + Bel (–A) = 1;

(g) Ignorance = 1 – {Bel (A) + Bel (–A);

(h) Plausibility (A) = 1 – Bel (–A).

Beliefs are combined according to Dempster's rule of combination, according to the following rules:

$$-\mathrm{Bel} = \mathrm{Bel}_1 \oplus \mathrm{Bel}_2 \qquad (1)$$

where $\oplus$ denotes the combination.

Under this rule, the mass committed to A, a subset under Bel, is

$$m(a) = \frac{\sum\limits_{A_i \cap B_i = A} m_1(A_i)\, m_2(B_j)}{1 - \sum\limits_{A_i \cap B_j = \phi} m_1(A_i)\, m_2(B_j)} \qquad (2)$$

where m_1 and m_2 is the mass committed to A or ϕ from different sources of evidence.

— Dempster's rule only works on a pair of belief function;
— The belief functions must be independent;
— The belief functions must refer to the same frame of discernment θ.

See figure 1 for an example of this rule; for another example for remote sensing applications see Srinivasan and Richards (1990).

As is discussed in Srinivasan and Richards (1990), a direct application of all aspects of Dempster's rule is complicated, because all labels within the frame of discernment have to be compared with one another. To make it work, a hierarchy of subsets has to be made or restrictions have to be made to make the method of practical use (Shafer and Logan, 1987).

2. Methodology used

In this study the following restrictions were made:

(a) Bel (θ) is assumed to be always 0.9;

(b) Only the belief against a particular label is expressed; consequently, only the plausibility of each label can be calculated;

(c) In this case, $m(-A) = (Bel(\theta))^2 - (m_1(\theta) - m_1(-A))* (m_2(\theta) - m_2(-A))$, so no normalization is necessary.

By expressing only the belief against a label, the need to compare all possible combinations of labels or the design of a hierarchical structure is not necessary. See figure 1 in this context; after a initial belief against A of 0.3 and 0.5 of two sources of evidence, the final mass of the belief against A is 0.81 – 0.24 = 0.57.

The final label for a pixel is classified after all the evidence is used, according to the following rules: the label is picked when the label with the least belief against that label is found; if two or more labels have the same amount of minimum belief against them, one of them is chosen. After the classification is complete, the results can be evaluated by displaying the level of evidence of a pixel.

This method is very flexible because missing information can be easily dealt with. For instance, even if for an area three images are available but one of the images is partly covered with clouds, the whole image is classified; in the areas with clouds, evidence from two images is used, and all other areas will use the complete set of data.

C. Knowledge base

Each pixel is assumed to have one of the following labels: rice, maize 1, maize 2, rice + second irrigated crop, rice + second non-irrigated crop, homestead garden/urban area, natural vegetation, or tamarind orchard. In this case study there are three sets of evidence to help attribute each pixel with a label: (a) the soil map 1:50,000 (LDD, 1990); (b) the Thematic Mapper (TM) image of December 1988; and (c) the TM image of March 1989.

1. Reflection of vegetation/soil system

The reflection of vegetation soil complex has been modeled and verified with the results of ground observations by various authors, such as Malila et al. (1977), Richardson and Wiegand (1990), and Major et al. (1990), to name a few. With the results of these studies estimations can be made of vegetation and soil characteristics on the bases of the reflection data in the near-infra-red (NIR) and red bands. The following observations can be made: (a) the amount of green vegetation can be estimated (GLAI); (b) with dense green vegetation cover, the internal and external shadow can be determined, an indication of the structure of the vegetation. However, it cannot be inferred from the relfection if the shadow is caused by shadows from the position of the leaves, or because of shadow caused by height differences between plants; (c) with low coverage of greeen vegetation, the brightness of the soil can be determined, although the amount of reflection sensed by the sensor is influenced not only by the amount of reflection of the soil but also by shadow effects caused by irregular surface features; (d) with intermediate vegetation coverage, the amount of shadow within the vegetation or the colour of the soil cannot be determined with high certainty. In areas with a large height difference, as well as with images with clouds, external shadow effects can play a role.

Also, differences in sun angle and recording angle between recording dates influence the amount of reflection. Several authors have tried to correct for these differences e.g. Richter (1990), Deering and Eck (1987), Crippen (1987), Hall et al. (1991), Chavez (1989), Ranson et al. (1986) and Singh (1988).

For an operational system using more satellite images, it would be useful to execute a such a radiometric correction on each image in order to get an estimation of the absolute reflection data, instead of the digital numbers (see Richter, 1990). This would help one infer the characteristics of the vegetation from the reflection data. In this study the images were only made comparable to cach other by overlaying the soil line and the vegetation line of the March image onto the corresponding lines of the December image (see Bronsveld, 1981, for the methodology).

2. Crop calendars

Wanjura and Harfield (1988) studied the differences in characteristics of several crops at difference stages of the growing season and the resulting reflection characteristics. In Cihlar et al. (1987), procedures are proposed to describe agricultural crops and soils in order to characterizing them for remote sensing studies. To a large extent they coincide with the information detectable from images, as described above. From those studies it becomes clear that the amount and the structure of green vegetation needs to be known at least at the time of the recordings of the images. If these characteristics are not observed they can be inferred by using information from cropping calendars. These give information about the time of plowing, planting, weeding and so on, together with climatic information, giving information of the amont of growth of crops after planting, or regrowth after harvesting.

3. Soil map

On the basis of the soil map, produced by the Land Development Department of Thailand, a land evaluation was execulted, using the ALES land evaluation expert system (see Rossiter and Van Wambeke, 1989). This produced information indicating the suitability for rice production. Also, information regarding the urban areas and water bodies, indicated on the map, was used.

D. Rule base

As each set of evidence relates to one source only, and the rules are dependent only on that particular set of evidence, the condition that the belief functions that are combined are independent is fulfilled.

1. Thematic mapper data

Both TM images were made comparable (see section C) and all possible combinations of values in bands 3 and 4 were classified according to a box classifier, as shown in figure 2. With that classifier the reflection was "translated" to cover classes. For each season and each cover class the belief against each label is estimated (table 1), making use of the knowledge of the amount and structure of the plant material and soil (see section C).

2. Soil maps

The rules associated with the soil map are based on the authors' personal knowledge of the study area, gathered by observations during fieldwork and the information from several reports of the area (Kuneepong, 1990). The rules, as shown in table 2, are based on the following knowledge: if soils are to some extent suitable for rice they will be used for rice, and not for maize. For security reasons, the farmers will always prefer rice (with or without a seond crop) above maize (rice is the main food crop, and maize is used as a cash crop). If soils are suitable for rice they will normally not be used for homestead gardens or for tamarind orchards. But soils suitable for rice (in the river plain) are usually associated with levees, which are used for homestead gardens, and stay in the class "very seldom unused-used" (natural vegetation). Soils which are indi-

Table 1a. The cover class of TM images with the assigned beliefs against each label if in the December image a pixel is attributed with that cover class

Code	Cover class	ma	m_2	ri	rm	nv	hg	ta	ir	wa
1.	Dark vegt.	0.8	0.8	0.8	0.8	0.1	0.1	0.8	0.8	0.8
2.	Mid. _vegt.	0.8	0.7	0.8	0.1	0.1	0.8	0.8	0.8	0.8
3.	Light_vegt.	0.8	0.7	0.8	0.8	0.1	0.8	0.8	0.8	0.8
4.	D. vegt./soil	0.8	0.8	0.8	0.8	0.2	0.5	0.8	0.8	0.8
5.	M. vegt./soil	0.8	0.2	0.8	0.8	0.2	0.1	0.4	0.8	0.8
6.	L. vegt./soil	0.8	0.2	0.8	0.8	0.2	0.8	0.8	0.8	0.8
7.	D. soil/vegt.	0.7	0.8	0.7	0.5	0.8	0.2	0.1	0.7	0.8
8.	M. soil/vegt.	0.2	0.6	0.7	0.2	0.5	0.2	0.6	0.7	0.8
9.	L. soil/vegt.	0.4	0.3	0.7	0.1	0.7	0.8	0.8	0.7	0.8
10.	Dark_soil	0.7	0.8	0.7	0.8	0.8	0.5	0.8	0.8	0.7
11.	Mid._soil	0.1	0.8	0.3	0.7	0.8	0.8	0.8	0.6	0.8
12.	Light_soil.	0.4	0.8	0.1	0.7	0.8	0.8	0.8	0.1	0.8
13.	Sl. scene_vegt.	0.8	0.1	0.8	0.8	0.7	0.8	0.8	0.8	0.8
14.	Mod. scene. vegt.	0.8	0.1	0.8	0.4	0.8	0.8	0.8	0.8	0.8
15.	Scene_vegt.	0.8	0.1	0.8	0.1	0.8	0.8	0.8	0.7	0.8
16.	Dead._vegt.	0.7	0.1	0.2	0.1	0.8	0.8	0.8	0.6	0.8
17.	Water	0.8	0.8	0.8	0.8	0.8	0.8	0.8	0.8	0.1
18.	Water/shade	0.8	0.8	0.8	0.8	0.1	0.8	0.8	0.8	0.5

Notes:
ri = Rice
ma = Maize 1
m_2 = Maize 2
ir = Rice + second irrigated crop
rm = Rice + second non-irrigated crop
hg = Homestead garden/urban area
nv = Natural vegetation
wa = Water body
ta = Tamarind

Table 1b. The cover class of TM images with the assigned beliefs against each label if in the March image a pixel is attributed with that cover class

Code	Cover class	ma	m_2	ri	rm	nv	hg	ta	ir	wa
1.	Dark vegt.	0.8	0.8	0.8	0.8	0.1	0.7	0.8	0.8	0.8
2.	Mid. _vegt.	0.8	0.8	0.8	0.8	0.1	0.7	0.8	0.7	0.8
3.	Light_vegt.	0.8	0.8	0.8	0.8	0.1	0.8	0.8	0.7	0.8
4.	D. vegt./soil	0.8	0.8	0.8	0.8	0.2	0.7	0.8	0.8	0.8
5.	M. vegt./soil	0.8	0.8	0.8	0.8	0.2	0.1	0.6	0.6	0.8
6.	L. vegt./soil	0.8	0.8	0.8	0.8	0.2	0.7	0.6	0.4	0.8
7.	D. soil/vegt.	0.6	0.6	0.8	0.8	0.2	0.7	0.5	0.8	0.8
8.	M. soil/vegt.	0.6	0.6	0.7	0.7	0.2	0.1	0.1	0.7	0.8
9.	L. soil/vegt.	0.8	0.7	0.7	0.7	0.8	0.7	0.8	0.7	0.8
10.	Dark_soil	0.8	0.6	0.8	0.8	0.8	0.8	0.8	0.8	0.7
11.	Mid._soil	0.1	0.1	0.3	0.3	0.8	0.8	0.8	0.8	0.8
12.	Light_soil.	0.3	0.3	0.1	0.1	0.8	0.8	0.8	0.8	0.8
13.	Sl. scene_vegt.	0.8	0.8	0.8	0.8	0.7	0.8	0.8	0.1	0.8
14.	Mod. scene. vegt.	0.8	0.8	0.8	0.8	0.8	0.8	0.8	0.7	0.8
15.	Scene_vegt.	0.8	0.8	0.8	0.8	0.8	0.8	0.8	0.8	0.8
16.	Dead._vegt.	0.8	0.7	0.5	0.5	0.8	0.8	0.8	0.8	0.8
17.	Water	0.8	0.8	0.8	0.8	0.8	0.8	0.8	0.8	0.1
18.	Water/shade	0.8	0.8	0.8	0.8	0.1	0.8	0.8	0.8	0.5

cated to be unsuitable-suitable for rice on the soil map might locally be suitable (in valleys too small to be mapped). The chance to find homestead gardens in this areas is very small. Often natural vegetation is often found here as well as tamarind orchards. On steeper land the chance to find rice is smaller than in the flatter areas. Urban areas are nearly exclusively used for homestead gardens.

Table 2. The data derived from the soil map with the assigned beliefs against each label, if a pixel is attributed with soil unit class

	ri	ma	m_2	rm	nv	hg	ta	ir	wa
S2 + 3	0.1	0.7	0.7	0.1	0.7	0.5	0.6	0.1	0.8
Flat S4	0.7	0.1	0.1	0.7	0.1	0.8	0.1	0.8	0.8
Not flat S4	0.8	0.1	0.1	0.8	0.4	0.8	0.1	0.8	0.8
Water	0.7	0.7	0.7	0.7	0.7	0.6	0.8	0.7	0.1
Urban	0.7	0.7	0.7	0.7	0.7	0.1	0.8	0.7	0.3
Outside	0.8	0.8	0.8	0.8	0.8	0.8	0.8	0.8	0.8

Note: S2, S3 and S4 refer to respond moderately suitable, marginally suitable and unsuitable for rice cultivation. "Flat" refers to flat and gently undulating areas, and "not flat" to undulating and steeper lands. "Water" refers to water bodies on the soil map. "Urban" refers to areas on the soil map. "Outside" is outside the study areas.

3. Maximum plausibility map

The evidence of the two satellite maps (belief against each label) is first combined according to the method described in section B.1, and the resulting map is then combined with the evidence of the soil map. As is described, for each pixel the label is chosen with the least evidence against it, in other words with the highest plausibility for a label compared with the other possible labels. Therefore, the resulting map will be called the maximum plausibility map (MP map) (see figure 3a). An MP map was produced, in which the level of the maximum plausibility is indicated. The lower the plausibility the more conflicting evidence was present, and the higher the chance of errors (figure 3b).

E. Results

The MP map was compared with the land-use map produced by LDD. The units of this map indicate the main crop(s) found within the units. The comparison was done because the field work was executed in 1985, some ten years after the aerial photographs were taken. In an area with rapid changes in land use, outdated photographs make it very time-consuming to produce maps that are accurate in the geo-referencing of land-use types or in the percentages of different land use within the unit. However, the comparison is useful because the change in land use of a certain area is limited to a few options. For instance, if the unit of the map was natural vegetation, it might have changed to maize, but not to rice. Rice areas will not be changed to maize. Only the first crop of the growing season is indicated on the map (nearly always rice). Tamarind orchards were not indicated on the map within the study area, because at the time of the photographs the number of tamarind orchards was limited. Even now the surface area of tamarind is not high, although it is rapidly growing.

The confusion matrix is presented in table 3. Because maize and natural vegetation are often in complex mapping units on the land-use map, maize and natural vegetation are often confused. Some of the mis-classifications evident in the matrix are caused by the differences in geo-referencing of the different maps. This is most evident in the pixels classfied as mixed orchards in the MP map (only 28 per cent are class-fied correctly). The units of this land use are typically long and narrow, making them very susceptible to errors. Units described in the land-use map as exclusively natural vegetation are in the MP. The map is simplified to map "other" as maize; this can be explained by the fact that the forests have four labels (land uses), increasingly invaded for maize cultivation. Pixels classified as rice in the MP map are often (21 per cent) maize on the land-unit map; this is mainly because of the fact that the suitability for rice is overestimated in two areas in the south-east and south-west parts of the study area. This should be corrected in a later version of the map. The mixed orchards are in only 28 per cent of the cases classified correctly. Pixels classified as tamarind orchards are frequently (61 per cent) found in areas with maize cultivation on the land-use map. This is as expected because maize fields are often changed to

Table 3. Confusion matrix

	Maize/natural vegetation	Rice	Natural vegetation	Mixed orchards
Maize	83,627 (71%)*	7,553 (6%)	25,236 (21%)*	12,839 (1%)
Rice	11,172 (21%)*	36,172 (69%)*	1,219 (2%)	2,560 (5%)
Rice + second crop	545 (7%)	6,856 (82%)*	– –	948 (11%)
Natural vegetation	87,009 (55%)*	2,459 (2%)	66,534 (42%)*	793 (1%)
Mixed orchards	6,389 (32%)	7,211 (37%)*	205 (1%)*	5,590 (28%)*
Tamarind orchards	3,537 (61%)*	907 (16%)	955 (7%)*	320 (6%)

Note: Horizontal figures are from the land-use map (from aerial photographs of 1974), and vertical figures are from the land-use map produced with the described method. The numbers represent the number of pixels classified. Please note that classes with vary low numbers of pixels in both maps are not displayed. Boxes indicated with * are probably carrectly classified (see figure 3c).

tamarind orchards. On the other hand, there is also much (secondary) natural vegetation near maize cultivation areas. As the reflections of tamarind orchards and natural vegetation are similar, this result should be regarded with caution.

A map was produced to indicate the pixels that were judged to be correctly classified (see figure 3c). The similarity between figures 3b and 3c indicates that low plausibility has a higher chance of mis-classification.

F. Conclusion

This study demonstrates that an integrated information system that includes GIS and remote sensing capabilities (the ILWIS/GIS software integrates GIS, remote sensing, relational database, and modelling techniques) is a powerful tool for the processing of satellite data and combining the data with information from other sources. It shows that such an information system makes it possible to update and monitor the present land cover and land use.

The method presented here will probably be an improvement on existing methods to process satellite images for land use mapping, because the amount of field work can be reduced and because of the improved flexibility to combine other sources of information with reflection data from satellites. The method will also be very flexible in cases of missing data. Although in this study, the results of the reported method looked very promising, not very much geo-referenced ground observation was done. Further fieldwork has to be done to decide whether the accuracy of this method is comparable to that of the methods like maximum likelihood classfication.

References

Anaman, T. (1990). Database development and use as a tool for farming systems analysis: a case study in Lom Kao and Na Saeng, upper Pa Sak watershed area, Phetchabun Province, Thailand. Unpublished MSc thesis. ITC, Enschede, Netherlands.

Bronsveld, M.C. (1981). Multitemporal Landsat imagery as a tool in land-use survey. In *Proceedings* of Landsat 81, Canberra, Australia.

Bronsveld, M.C., and F.J.D. Luderius (1982). Analysis of multitemporal data for the identification of crops. ITC, Enschede, Netherlands.

Chavez, Pat S., Jr. (1989). Radiometric calibration of Landsat Thematic Mapper multispectral images. *Photogrammetric Engineering and Remote Sensing,* 55(9): 1,285-1,294.

Cihlar, J., M.C. Dobson, T. Schmugge, P. Hoogeboom, A.R.P. Janse, F. Baret, G. Guyot, T. Le Toan and P. Pampaloni (1987). Review article: Procedures for the description of agricultural crops and soils in optical and microwave remote sensing studies. *International Journal of Remote Sensing,* 8(3): 427-439.

Crippen, Robert E. (1987). The regression intersection method of adjusting image data for band ratioing. *International Journal of Remote Sensing,* 8(6): 893-916.

Deering, Donald W., and Thomas F. Eck (1987). Atmospheric optical depth effects on angular anisotropy of plant canopy reflectance. *International Journal of Remote Sensing,* 8(3): 427-439.

Gangkofner U., A. Relin and S. Saradeth (1990). Regional inventory 1988-1989 in Bavaria. In *Proceedings* of 1989 Conference on Application of Remote Sensing to Agricultural Statistics. EEC Publications.

Hall, F.G., D.E. Strebel, J.E. Nickeson and S.J. Goetz (1991). Radiometric rectification: toward a common radiometric response among multi-date, multi-sensor images. *Remote Sensing of the Environment,* 35: 11-27.

Huizing, H., and M.C. Bronsveld (1991). The use of geo-information systems and remote sensing for evaluating the sustainability of land-use systems. Paper presented at International Workshop of Evaluation for Sustainable Land Management, Chiang Rai, Thailand.

Kuneepong, P. (1990). Crop modelling of maize as a tool on the database and as an input to land evaluation for the upper Pa Sak watershed, Pa Sak watershed, Phetchabun, Thailand. Unpublished MSc thesis. ITC, Enschede, Netherlands.

Major, D.J., F. Baret and G. Guyor (1990). A ratio vegetation index adjusted for soil brightness. *International Journal of Remote Sensing,* 11(5): 727-740.

Malila, W.A., J.M. Gleason and R.C. Cicone (1977). Multispectral system analysis through modelling and simultation. In *Proceedings* of International Symposium of Remote Sensing on Environment, Ann Arbor, Michigan.

Ranson K.J., C.S.T. Daughtry and L.L. Biehl (1986). Sun angle, view angle, and background effects on spectral response of simulated balsam fir canopies. *Photogrammetric Engineering and Remote Sensing,* 52(5): 649-658.

Richardson, Arthur J., and Craig L. Wiegand (1990). Comparison of two models for simulating the soil-vegetation composite reflectance of a developing cotton canopy. *International Journal of Remote Sensing,* 11(3): 447-459.

Richter, Rudolf (1990). A fast atmospheric correction algorithm applied to Landsat TM images. *International Journal of Remote Sensing,* 11(1): 159-166.

Rossiter, D.G., and A.R. Van Wambeke (1989). *Automated Land Evaluation System (ALES): Version 2 User's Manual.* Ithaca, New York: Department of Agronomy, Cornell University.

Shafer, G., and R. Logan (1987). Implementing Dempster's rule for hierarchical evidence. *Artificial Intelligence,* 33: 271-298.

Singh, S.M. (1988). Simulation of solar zenith angle effect on global vegetation index (GVI) data. *International Journal of Remote Sensing,* 19(2): 237-248.

Srinivasan, A., and Richards (1990). Knowledge-based techniques for multi-source classification. *International Journal of Remote Sensing,* 11(1): 505-524.

Wanjura, D.F., and J.L. Hatfield (1988). Vegetative and optical characteristics of four-row crop canopies. *International Journal of Remote Sensing,* 9(12): 249-258.

$m_1(-A) = 0.5, m_1(\theta) = 0.9$
$m_2(-A) = 0.3, m_2(\theta) = 0.9$

	$m_1(-A)$	$m_1(-A) - m_1(\theta)$
$m_2(-A)$	$m(-A) = (0.5 \times 0.3) =$ 0.15	$m(-A) = (0.4 \times 0.3) =$ 0.12
$m_2(\theta) - m_2(-A)$	$m(-A) = (0.5 \times 0.6) =$ 0.30	$m(\theta) = (0.6 \times 0.4) =$ 0.24

Figure 1. Combining evidence using Dempster's rule (modified from Srinivasan and Richards, 1990)

TM 4

3	6	13	13	13	14	14	15	15	16
3	6	6	13	13	14	14	15	15	16
3	6	6	6	6	9	9	15	16	16
2	2	5	5	8	9	9	9	12	12
2	5	5	5	8	8	12	12	12	12
2	5	5	8	8	11	12	12		
1	4	7	7	11	11				
1	4	7	10	11					
1	10	10	10						
18	17	17	17						

TM 3

Figure 2. The feature space of TM bands 3 and 4. The numbers indicate the cover class code. For the class description see table 1

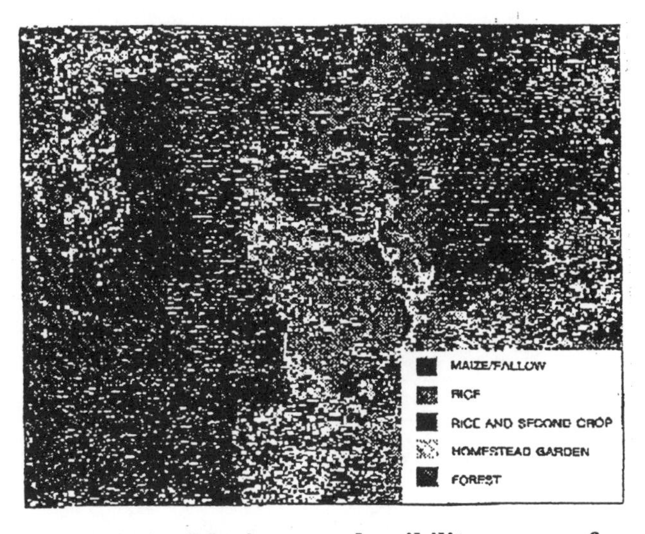

Figure 3(a). Maximum plausibility map; for each pixel a label is chosen with the least evidence against it. The map is simplified to only four labels (land use)

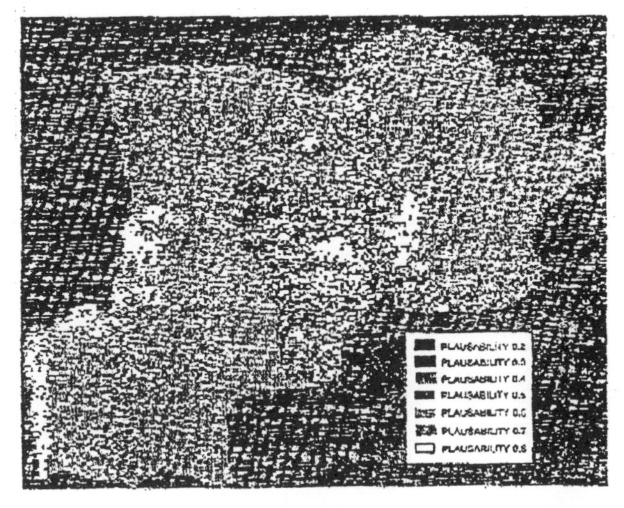

Figure 3(b). Map indicating the maximum value of plausibility of each pixel

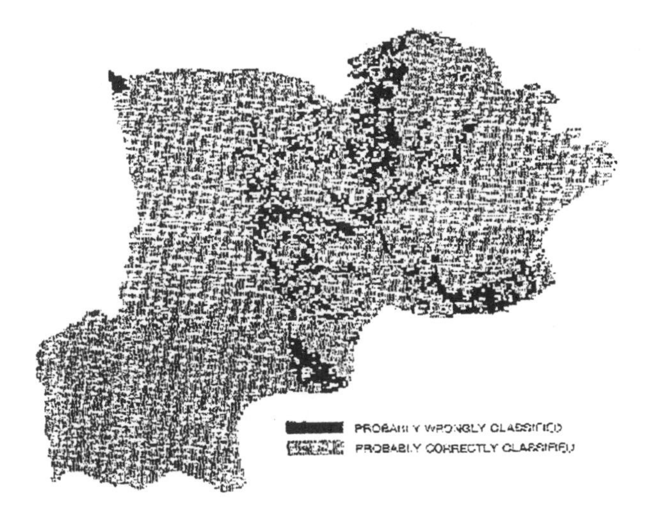

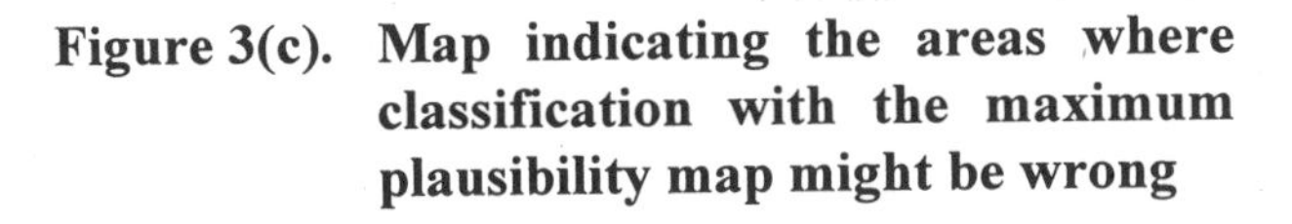

Figure 3(c). Map indicating the areas where classification with the maximum plausibility map might be wrong

PART FIVE
AGRICULTURE AND SOILS

REMOTE SENSING APPLICATIONS IN AGRICULTURE

*M.V. Krishna Rao**

A. Introduction

Agricultural resources are important renewable dynamic natural resources. In India, the agricultural sector alone provides livelihood for about 70 per cent of the population and forms the backbone of the economy. Increasing agricultural productivity has been the main concern to meet the ever-increasing demands on food production due to the alarming rate of population growth. In this context, comprehensive and reliable information on soils, extent of wastelands, agricultural crops and water resources, both surface and underground, is most useful for optimal management of agricultural resources. Information on crops, their acreage, condition and production enables the planners to adopt suitable measures to meet shortages, if any, and implement proper support and procurement policies. Remote sensing technology has opened new vistas by virtue of its ability to provide regular, synoptic coverage of the country at periodic intervals, and it is playing an important role in providing such information.

B. Remote sensing

The term "remote sensing" may be defined as the science and art of acquiring information about objects from measurements made from a distance, without coming into physical contact with the object. The human eye, cameras, and multispectral scanners are some examples of remote sensors. Remote sensing is largely concerned with the electromagnetic energy that is scattered or emitted by objects receiving energy from sun or any other external source. Different objects return different amounts of electromagnetic energy, which is reflected in different bands of the electromagnetic spectrum, and these differences, in turn, are helpful for identification of various ground objects. Today, remote sensing data are available from different platforms, like ground-based radiometers, low- and high-altitude aircraft and space-borne satellites. Remotely sensed data can be obtained either as photographic images, which are interpreted visually by photo-interpretation techniques, or as digital data on computer compatible tape (CCT), which are analysed using suitable computers.

Remote sensing technology has the potential to provide information about various agricultural resources that may have direct or indirect bearings on food production. For example, information about the extent, distribution and condition of crops -- coupled with information about soils and associated problems, water resources for agriculture, impact of land use on agriculture and so forth -- can be collected and integrated using various remote sensing techniques. The basic information that remote sensing techniques (satellite or airborne) are expected to contribute to agricultural crop studies are data related to (a) crop identification, (b) crop acreage estimation, (c) crop condition assessment and (d) crop yield forecasting. Crop identification has long been recognized as one of the potential applications of remote sensing. Crop identification and acreage estimation of crops grown in a region form the basic information not only for assessing crop yield potentials but also for agricultural planning, export and import negotiations, storage, pricing and distribution of agricultural commodities.

C. Indian experience

In India, identification of coconut wilt disease in 1970 using aerial colour infra-red photographs was the pioneering experiment (Dakshinamurthy et al., 1971). This was followed by the Agricultural Resources Inventory Survey Experiment (ARISE), carried out jointly by the Indian Space

* Agriculture Division, National Remote Sensing Agency, Hyderabad, Andhar Pradesh, India.

Research Organization and the Indian Council of Agricultural Research in Anantapur District of Andhra Pradesh and in Patiala District of Punjab, to estimate acreage under various crops using aerial colour infra-red photographs. Spectral reflectance information collected with an 11-band airborne Bendix multispectral scanner by the National Remote Sensing Agency (NRSA) demonstrated its potential utility to identify different stages of paddy crop (Krishna Rao et al., 1982; Ayyangar et al., 1980). In the field experiments, significant correlations were observed between leaf area index (LAI), yield and spectral reflectance of the paddy crop.

Under the Indian Remote Sensing Satellite-Utilization Programme (IRS-UP), initiated by the Department of Space (DOS) in 1984, different aspects of the problem of crop inventory using remotely sensed data were studied. Systematic plot-level experiments were carried out on wheat and rice crops to understand their spectral behaviour and to develop empirical yield models using spectral data. Concepts developed have since been used to estimate yield using space-borne data. The first attempt in the country to use satellite digital data for wheat acreage estimation was made in Karnal District of Haryana by the Space Applications Centre (SAC), using Landsat MSS data (Dadhwal and Parihar, 1985).

The Crop Acreage and Production Estimation (CAPE) project, under the Remote Sensing Applications Mission, with enlarged scope and objectives, was formulated in 1985, to cover all the major food, oil-seed and fibre crops of India. A concerted effort has been made under this programme to develop an automated method for reducing the turn-around time to obtain the in-season pre-harvest crop estimates (Sahai, 1990; Navalgund, 1991).

D. Acreage estimation

Identification and discrimination of various crops and land-cover classes require quantitative use of subtle differences in their spectral data and digital image processing techniques. The acreage estimation procedure broadly consists of identifying representative sites of various crops and land-cover classes in the image based on the ground truth collected, generation of signatures for different training sites and classification of the image using training statistics.

The majority of the work carried out so far has used single-date data corresponding to the near maximum vegetative growth stage of the crop. District level and sub-district level acreage estimates are generally reached by use of the total enumeration approach. In this case, the administrative boundary of the district or the sub-district unit is digitized, and then a mask is generated and overlaid on the scene. All the data elements (pixels) composing the mask are classified by employing a supervised maximum likelihood procedure.

Estimation of crop acreage for large areas like districts and of states requires handling of a very large volume of data, greater efforts in ground truth data collection, and similar considerations. In such cases, it will be difficult to complete the entire data analysis in the short time available for providing pre-harvest estimates. To overcome this problem, sampling technique-based procedures have been developed and are successfully being used. Based on the crop concentration statistics, agro-physical and/or agro-climatic conditions, the study area is first divided into homogeneous strata. Each stratum is further subdivided into segments, each of 10 km x 10 km size. Segment sizes of 7.5 km x 7.5 km and 5 km x 5 km are used for heterogeneous areas. Further stratification is based on the agricultural area of the sample segment. Ten per cent of the total population segments are used in sampling for extraction of area statistics. Appropriate statistical methods are employed to aggregate results, and thus crop acreage figures at study-area and state levels are obtained. These procedures have been successfully operationalized under the CAPE project.

1. Mono-cropped areas

In Andhra Pradesh, district-wise acreage and production estimates are being generated for the major crops -- paddy (kharif and rabi), sorghum (rabi), groundnut, cotton and mesta -- which occupy at least 5 per cent of the geographical area of the district. Cotton is the most important fibre cash crop of Andhra Pradesh. The acreage and production of this crop are being estimated for seven districts of the state, adopting the stratified random sampling approach (Venkataratnam et al., 1993a). Soybean is another important oil-seed cash crop of Madhya Pradesh. During the 1994-1995 kharif (monsoon) season, cropped area statistics of seven districts of Madhya Pradesh were generated using remote sensing techniques (Venkataratnam et al., 1993b).

The relative bias and the coefficient of variation, which respectively denote the accuracy and reliability of the results, were calculated to test wheter these estimates met the 90/90 accuracy/confidence criterion. In both cases, the remote sensing estimates satisfied the 90/90 criterion.

2. Multi-cropped areas

The remote sensing technology proved its utility in mono-cropped areas of large, contiguous and homogeneous nature. However, studies using satellite remote sensing data under heterogeneous and multiple cropping situations are limited. Fortunately, the availability of finer spatial resolution (23.5 m) of LISS-III data of IRS-1C provides opportunities to derive information on crops growing in small holdings under multiple cropping situations. In a study conducted by NRSA, using LISS-III data of January for two different test sites -- Amaravati and Prattipadu villages of Guntur District, Andhra Pradesh, each with a total cropped area of about 10,000 ha -- up to six crop types, i.e. paddy, cotton, chillies, maize, bananas and tobacco, could be identified. These crops showed up distinctly on the LISS-III image and could be classified.

3. Horticultural crops

Horticulture plays an important role in an agricultural economy. Horticultural crops cover a wide array of crops such as fruits, vegetables, plantation crops, spices, condiments, medicinal and aromatic plants. Besides, the spectral reflectance pattern of these crops, by virtue of their pereniality, seasonal behaviour and specific geometry, are easy to discriminate from the field crops, which are seasonal in nature. However, wide plant spacing and intercropping within orchards interferes with the spectral reflectance of orchards. Certain pilot investigations were carried out at NRSA and other Department of Space (DOS) centres, to identify and estimate acreage of mango, cashew nut, coconut, banana and citrus orchards in different parts of the country, and encouraging results were obtained. A Ministry of Agriculture-funded project for coconut in Tamil Nadu, Karnataka and Kerala is in progress.

E. Microwave data

A lack of cloud-free data at the optimum growth periods during kharif season, the major crop growing season of India, is a major problem. Microwave remote sensing data, by virtue of its all-weather and cloud penetration capability, is the potential alternative to optical data as a source of space-borne satellite data. With the launch of microwave satellites with synthetic aperture radar (SAR) systems -- such as the ERS satellites with C band, W polarization; the JERS with L band, HH polarization; and the Radarsat with C band, HH polarization -- there are more opportunities for obtaining data for agricultural crop studies.

ERS-1 SAR data over two test sites, i.e. three successive individual and time composite FCC images, were analysed at NRSA through visual interpretation (Rao et al., 1994). False colour composites (FCCs) made out of SAR temporal data were found to have better crop discrimination than single-date data, because of the changes that took place with crops during the observation period, which were more clearly manifested on the FCC. Plantations of coconut and banana did not undergo major change in crop geometry or density and showed uniform DN values in all three passes. Paddy, cotton and tobacco showed changes in radar return with changes in crop growth stages during the observation period. However, no discrimination was observed between bare ground and pulse crops because of the near-transparency of pulse crops to C-band frequencies.

In another study area, it was possible to discriminate paddy and cotton crops using the filtered ERS-1 SAR data of October along with a corresponding texture image. Addition of the textural information with the filtered data was found to improve the separability of these two crops. These studies indicate the need for more work to be carried out on the use of SAR data on an operational basis for crop acreage estimation.

E. Yield estimation

Yield is influenced by a large number of factors such as crop genotype, soil characteristics, cultural practices adopted, weather conditions and biotic influences such as weeds, diseases and pests. Spectral information about a crop is the integrated manifestation of the effects of all these factors on its growth. The two approaches generally available for yield modelling using remote sensing data are (a) relating remote sensing data or derived parameters directly to yield and (b) relating remote sensing data to the biometric parameters, which in turn serve as input parameters in yield models. The spectral index of the crop canopy, expressed as ratio vegetation index (in the near-infra-red and red bands, NIR/R) or as normalized difference vegetation index [(NIR-R)/(NIR+R)] at any given time indicates the crop growth as affected by various factors in the time domain. Efforts have been made to develop yield-spectral index relationships using space-borne spectral indices measured at maximum vegetative cover and also using different parameters of the profile.

G. Crop monitoring and condition assessment

Monitoring and assessment of crop condition is feasible with the availability of multitemporal satellite data during the crop growth season. The daily NOAA-AVHRR at 1.1 km spatial resolution and the five-day repeating WiFS/IRS-1C at 188 m provide information on the crop condition at a regional level. Studies carried out with the early available WiFS data of January showed the potential utility of deriving regional-level estimates of paddy, cotton and pulse crops in Andhar Pradesh (Navalgund et al., 1996). Analysis of the WiFS data of the following February showed harvesting of pulse crops and standing paddy crop. Thus, the temporal WiFS data has been found to be useful in monitoring cropping patterns and systems.

Condition of the crop is affected by factors such as supply of water and nutrients, insect or pest attack, disease outbreak and weather conditions. These stresses cause physiological changes that alter the optical and thermal properties of leaves and bring about changes in canopy geometry and reflectance/emission. Effective crop condition assessment requires (a) detection of stress, (b) differentiation of stressed crop from the normal crop at a given time, (c) quantification of extent and severity of stress and (d) assessment of the production loss. Condition assessment is normally done on a grid-cell basis using multi-band satellite data. The area of interest is divided into geographically referenced grid cells of appropriate size, and each grid is monitored individually. The vegetation index (VI) for the pixels of crop of interest is computed for the selected sample segments and VI statistics are generated, which are eventually aggregated to compute the vegetation index number (VIN). A VIN profile is generated by computing VIN for different stages of the crop growth; the profiles are stored in a database to carry out trend analysis.

H. Conclusions

Remote sensing techniques have been operationally used in India to provide basic information not only on crops, but also on the impact of drought and flood on agriculture. Procedures for pre-harvest acreage estimation of major crops, using sampling and digital techniques, have been developed and successfully used every season, meeting the 90/90 accuracy/confidence criterion. Several efforts to improve crop classification accuracy and yield forecasts are in progress. Integrated studies on soil and water management using remote sensing and geographic information system techniques are also under active development to help achieve sustainable increases in agriculture production.

References

Ayyangar, R.S., M.V. Krishna Rao and K.R. Rao (1980). Interpretation and analysis of multispectral data for agricultural crop cover types: role of crop spectral response and crop spectral signatures. *Photonirvachak, Journal of Photo Interpretation and Remote Sensing*, 5: 39-48.

Dadhwal, V.K., and J.S. Parihar (1985). Estimation of 1983-1984 wheat acreage of Karnal District (Haryana) using Landsat MSS digital data. Technical note, IRSUP/SAC/CAPE/SR/25/90, Space Applications Centre, Ahmedabad, India, pp. 51-76.

Dakshinamurthy, C., B. Krishnamurthy, A.S. Summanwar, P. Santha and P.R. Pisharoty (1971). Remote sensing of coconut wilt. In *Proceedings* of the Sixth International Symposium on Remote Sensing of Environment, Ann Arbor, Michigan, pp. 25-29.

Krishna Rao, M.V., K.S. Ayyangar and P.P. Nageswara Rao (1982). Role of multispectral data in assessing crop management and crop yield. In *Proceedings* of the Eighth Machine Processing of Remotely Sensed Data Symposium, West Lafayette, Indiana, 7-9 July, pp. 226-233.

Navalgund, R.R. (1991). Remote sensing applications in agriculture: Indian experience. In *Proceedings* of Special Current Event Session, International Astronautical Federation, forty-second IAF Congress, Montreal, Quebec, pp. 31-50.

Navalgund, R.R., J.S. Parihar, L. Venkataratnam, M.V. Krishna Rao, S. Panigrahy, M.C. Chakraborty, K.R. Hebbar, M.P. Oza, S.A. Sharma, N. Bhagia and V.K. Dadhwal (1996). Early results from crop studies using IRS-1C data. *Current Science* (Special Section: IRS-1C), 70: 568-574.

Rao, P.V.N., K. Srinivas, K.V. Ramana, L. Venkataratnam and M.V. Krishna Rao (1994). ERS-1 SAR data for crop identification. In *Proceedings* of the National Symposium on Microwave Remote Sensing Users' Meeting, Space Applications Centre, Ahmedabad, 10-11 January, pp. 129-134.

Sahai, B. (1990). RSAM project -- Crop Acreage and Production Estimation (CAPE): an overview. Status report on "Crop acreage and production estimates". SAC report: RSAM/SAC/CAPE/SR/25/90, Space Applications Centre, Ahmedabad, India, pp. 1-22.

Venkataratnam, L., M.V. Krishna Rao, T. Ravi Sankar, S.S. Thammappa and S. Gopi (1993a). Cotton crop acreage/estimation and condition assessment through remote sensing method. In *Proceedings* of the National Seminar on Oil-seed Research and Development in India: Status and Strategies, pp. 217-218.

Venkataratnam, L., T. Ravi Sankar, V.L. Kumar and V.V.R. Babu (1993b). Pre-harvest acreage estimation of soybean using IRS data for forecasting production. In *Proceedings* of the National Seminar on Oil-seed Research and Development in India: Status and Strategies, pp. 214-216.

APPLICATION OF GIS FOR EVALUATING CULTIVATION AREA OF RED PEPPER IN THE REPUBLIC OF KOREA

*Keun-Seop Shim**

A. Introduction

Pepper is one of the important vegetables in the hot, spicy flavour of Korean kimchi. The total amount of red pepper produced in the Republic of Korea was 176,000 tons in 1995. The main factors affecting the red pepper yield are the temperature and sunshine hours in summer and the soil condition. But it is very difficult to evaluate the farming area by considering various environmental factors simultaneously, and the yield of pepper is relatively low in some places. Therefore, it would be useful if an evaluation system were developed to help in selecting the proper area for cultivating pepper. It is one of the purposes of this paper to introduce the possibility of using a geographic information system (GIS) in evaluating the cultivation area, because a GIS makes it easier to consider various kinds of environmental factors in combination.

B. The flow of the system

The system consists of three parts (figure 1). The first part is the regional database, which consists of environmental factors, such as weather and soil component. The second part is the evaluation table, in which are categorized the environmental factors; the ranges and scores were decided in discussions between several researchers. The last part is the evaluation programme, including the mapping procedure of the result.

C. Database and software

SAS/GIS and Arc Info programmes were used in evaluating the region. Map data was digitized from a map at 1:25,000 scale (figure 2). The soil database is composed of physical characteristics such as the nature of the soil, depth of the earth, draining conditions and geographical features; chemical characteristics include pH, K, Ca, OM (organic matter), Mg, CEC (cation exchange capacity) and SiO_2 (silica). Meteorological data consists of average temperatures, maximum or minimum temperatures, duration of sunshine, and precipitation.

The evaluation was conducted by summing the scores of each meteorological factor as categorized in table 1, and mapping was done by using the computer software. The scores of each soil factor are shown in table 2.

D. Results

The evaluation of red pepper suitability is classified as "very suitable", "moderately suitable" and "little suitable", depending on the weather and soil condition of a region. Anseong County of Kyeongki Province, which is in the central part of the Republic of Korea, was selected as the demonstration region to evaluate the system. The result of the classification, according to various kinds of environmental factors, showed different patterns, so it was very difficult to summarize. But six polygons in figure 3 (f), after consideration of soil and meteorological factors, proved to be moderately suitable areas, and the others were little suitable.

* Rural Development Administration, Suwon, Republic of Korea.

E. Conclusion

The application of GIS in agriculture is increasing rapidly in the Republic of Korea and is expected to provide various kinds of information to farmers, so that they can manage their lands much more effectively. But an effective GIS requires a large number of standardized databases and much information, which means that the researchers of various fields should work in collaboration. In this way, the evaluation system of crop suitability by region could be successful, once many research databases are integrated.

Table 1. Range and score of various meteorological factors

Meteorological factors	Range and score
	<15, 16~20, 21~25, 26~30, 31~35, 35> (2) (4) (6) (10) (2) (1)
Temperature (°C) minimum	>0, 1~5, 6~10, 11~15, 16~20, 21> (1) (2) (4) (6) (10) (6)
Average	>0, 1~5, 6~10, 11~15, 16~20, 21~25, 25> (1) (2) (4) (6) (8) (10) (6)
Precipitation (mm/month)	>50, 51~100, 101~150, 151~200, 201~250, 251~300, 300> (2) (8) (10) (6) (4) (2) (1)
Duration of sunshine (hour/month)	>100, 101~150, 150~200, 201~250, 250> (2) (4) (6) (8) (10)

Table 2. Range and score of soil chemical characteristics

Characteristics of soil	Range and score
pH	6.0~6.5, 5.5~6.0 or 6.6~7D, 5.4 or 7.0> 5.3, 5.2, 5.1< (10) (9) (8) (7) (6) (5)
Ca	5.0~6.0, 4.5~5.0 or 6>4.0~4.5, 4.5~3.0 (10) (9) (8) (6)
CEC	10~15, 9.2~9.9 or 15> 8.0~9.6, 8.0< (10) (9) (7) (5)

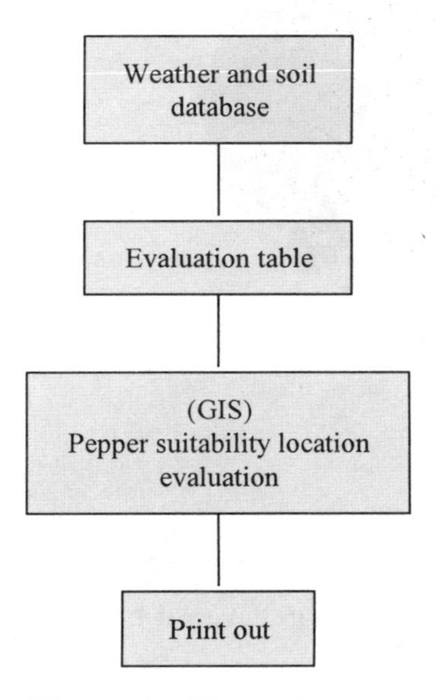

Figure 1. Flow of system

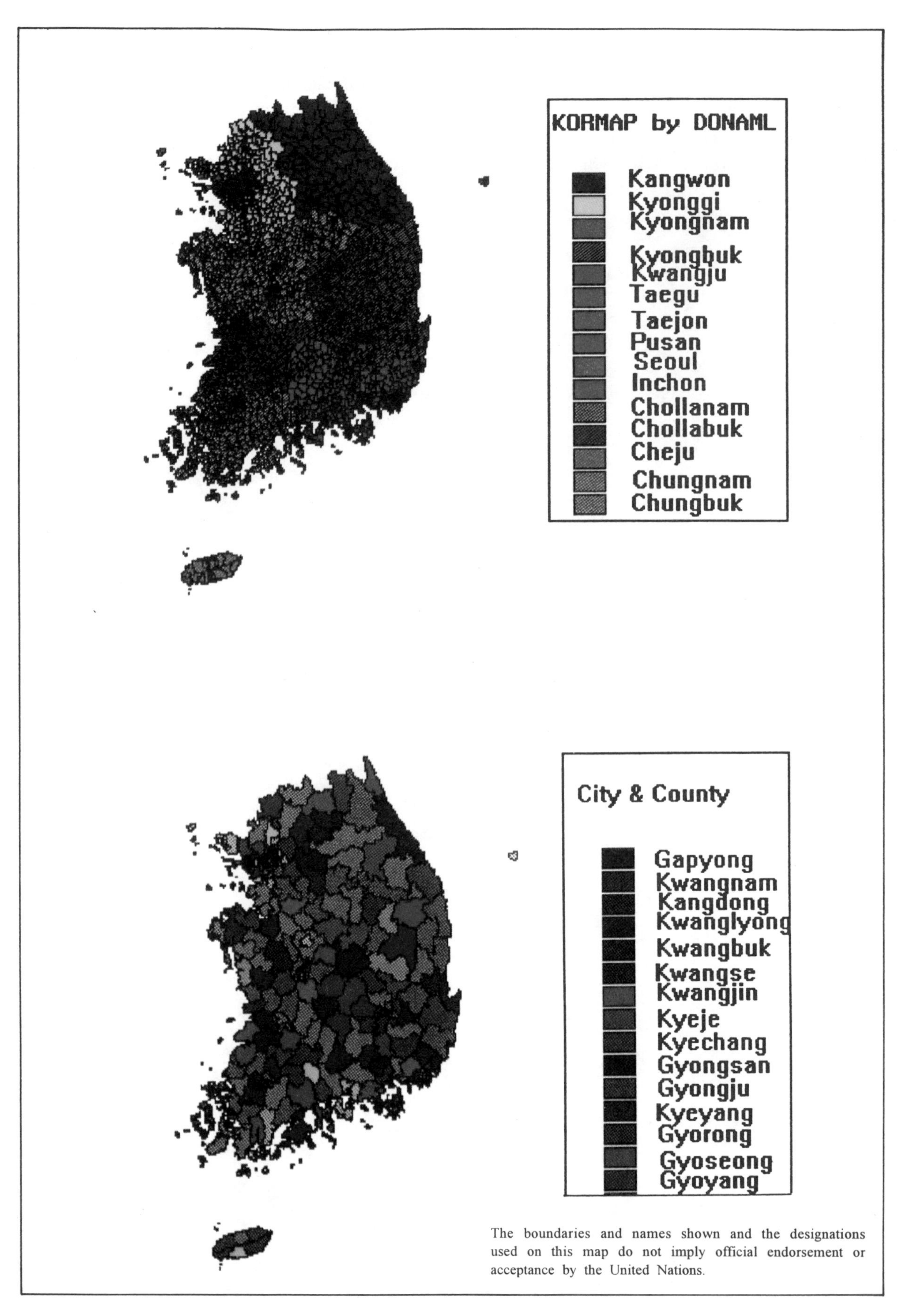

Figure 2. Map of the Republic of Korea generated by SAS/GIS

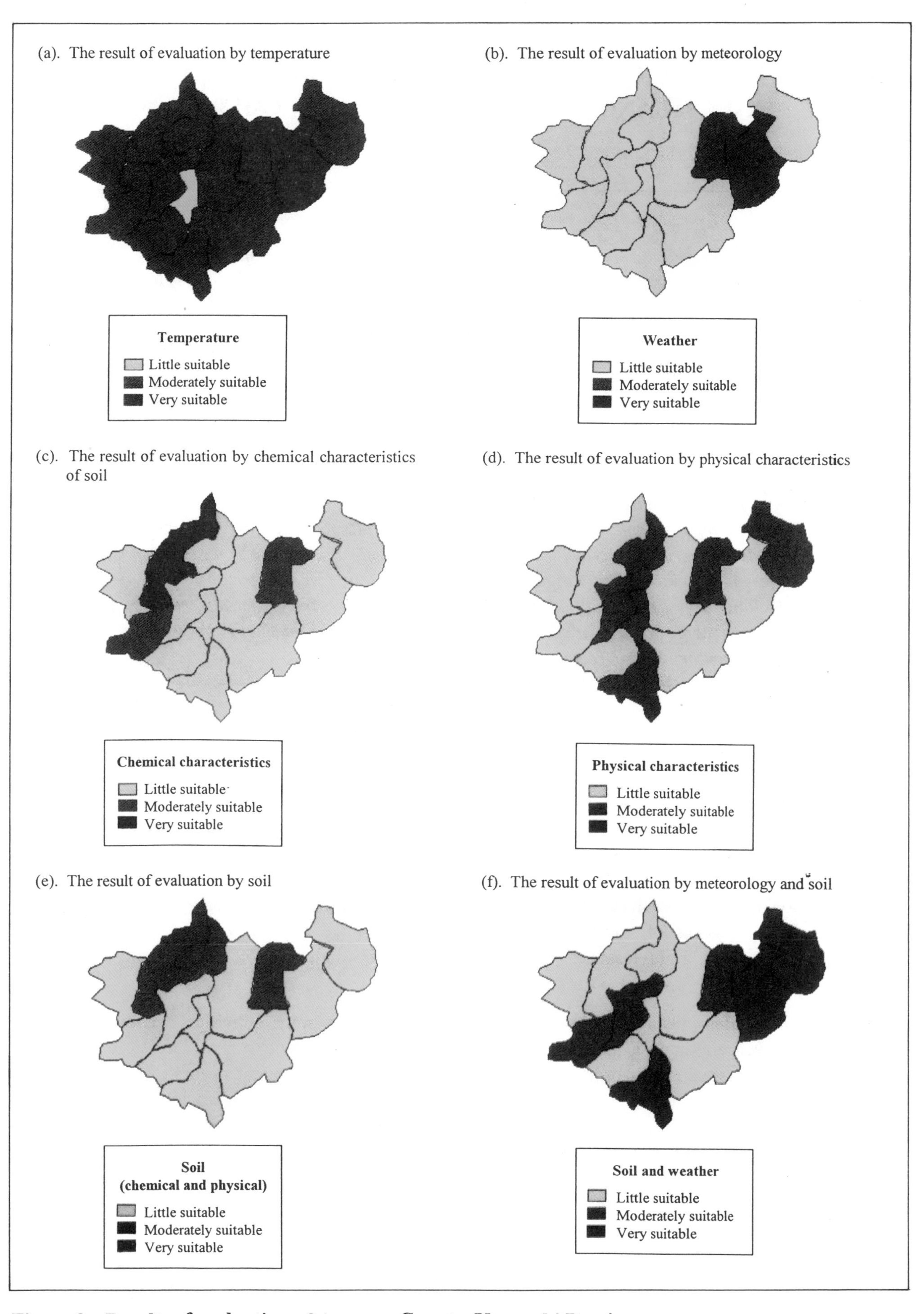

Figure 3. Results of evaluation of Anseong County, Kyeongki Province

WINTER WHEAT DROUGHT MONITORING BASED ON NOAA-AVHRR IN HUANG-HUAI-HAI REGION OF CHINA

*Tang Huajun and Xiao Shu**

ABSTRACT

The Department of Agricultural Application of the National Remote Sensing Centre of China conducts research in agro-disaster monitoring and crop acreage and yield estimates, as well as land use change studies. This paper presents a case study of the winter wheat drought monitoring in the Huang-Huai-Hai region of China.

Drought is the most important natural disaster for China's agricultural production, resulting in approximately five million tons of grain loss almost every year. An operational case study of winter wheat drought assessment and monitoring was undertaken in 1995 based on NOAA-AVHRR data in the Huang-Huai-Hai region. The method of Crop Water Stress Index (CWSI = 1-E/Ep) and five drought classes (severe, intermediate, light, none and moist) were used, based on the percentage of soil moisture content divided by soil field capacity of the upper 20 cm of soil. The results of both (a) percentage of winter wheat water stress acreage divided by sowing acreage and (b) soil moisture distribution maps for different provinces and counties per "decade" (that is, a period of ten days) from February to June are described. Although comparison of the research results and field data show 82 per cent accuracy, comparative studies of different methodologies for drought assessment and monitoring are recommended.

A. Introduction

About one third of the global land area is located in the arid and semi-arid regions. Drought is a severe climatic problem for agricultural production all over the world, which directly threatens socio-economic development. In China particularly, drought occurs more frequently, lasts longer and affects larger area than any other disaster. From the year 206 A.D. to 1949, there were 1,056 instances of severe drought, and from 1950 to 1995 there were 12 years in which more than 28 hectares million of arable land suffered from severe drought. About 20 million hectares of farmland per year, i.e. 20 per cent of the total arable land, was affected by severe drought, which caused about 33 million kg of grain loss per year. Drought loss accounts for 52 per cent of all disaster-related losses in China.

Agriculture suffers from drought most severely of all economic activities. Drought reduces yield and may even eliminate a harvest. Therefore, it is very important to find a feasible method to monitor the drought situation in order to take proper measures once drought occurs. With the advantage of large scales, remote sensing provides us with multispectral information about the land surface and is the most immediate and most important source of information with which to monitor drought. Researchers both at home and abroad have performed numerous studies on monitoring drought situations by remote sensing. This paper presents a case study of winter wheat drought monitoring in the Huang-Huai-Hai region of China.

* Department of Agricultural Application, National Remote Sensing Centre of China/Institute of Natural Resources and Regional Planning, Chinese Academy of Agricultural Sciences, Beijing.

B. Materials and methods

1. Materials

According to Tian (1990), there are five centres of drought in China: Huang-Huai-Hai drought zone, Southern Coast drought zone, South-Western drought zone, North-Eastern drought zone, and North-Western drought zone. Figure 1 shows the seasons and impacts of droughts in major areas of China.

The Huang-Huai-Hai drought zone is the most severely affected drought area among the five centres of drought in China. The annual rainfall is relatively low and varies greatly year to year. Therefore the Huang-Huai-Hai region is selected as the case study area. The location of the case study area is shown in figure 1.

The Huang-Huai-Hai (HHH) region is located in the centre of the Chinese Economic Development Zone. To the north is Beijing and Tianjin, to its south the Yangtze River Economic Zone, and on its east the Bohai Sea and Yellow Sea, where many ports such as Tianjin, Qinhuangdao, Qingdao, Yantai and Lianyungang are located. The HHH area is formed by deposition in the lower reaches of the Yellow River (Huang River), the Huai River, and the Hai River. It includes most part of the Hebei, Henan, and Shandong provinces and parts of Jiangsu and Anhui provinces. However, the HHH area defines one of the nine agricultural zones in China, according to Zhang's *Comprehensive Agricultural Regionalization of China.* There are 279 counties, 189 million people and 20 million hectares of arable land in the plain area. Most of the plain has flat topography, deep soil and proper soil texture, which are suitable for crop growth. The HHH plain represents one of the oldest cultivated lands on Earth. It produces more than 50 per cent of China's wheat. The rapid commercialization in southern China has increased China's dependency on a steady food supply from the plain and has made it the food basket of tomorrow's China.

2. Methods

There are many remote sensing methods developed for drought studies. In this paper, we use the Crop Water Stress Index (CWSI) method to study and monitor the drought situation during the winter wheat growing period.

The Crop Water Stress Index is obtained by applying an inferred model based on the principles of the Energy Balance-Crop Resistance Model to calculate evapotranspiration and potential evapotranspiration under the conditions of vegetation cover. CWSI is related to soil water content, so it is used as an indicator of soil moisture condition.

Crop Water Stress Index (CWSI)

$$= 1 - E/Ep \tag{1}$$

$$= \frac{\gamma^*\{1 + [\gamma_c/(\gamma_{ac} + \gamma_{bh})]\} - \gamma^*}{\Delta + \gamma^*[1 + \gamma_c/(\gamma_{ac} + \gamma_{bh})]} \tag{2}$$

where $\gamma^* = \gamma^*[1 + \gamma_{cp}/(\gamma_{ac} + \gamma_{bh})]$ (3)

$$\gamma_c(\gamma_{ac} + \gamma_{bh})] = \frac{\gamma(\gamma_{ac} + \gamma_{bh})\ (R_n - G)/(\rho C_p) - (T_s - T_a)\ (\Delta + \gamma) - (e_{a*} - e_a)}{\gamma[(T_s - T_a) - (\gamma_{ac} + \gamma_{bh})\ (R_n - G)/(\rho C_p)]} \tag{4}$$

with R_n = Net total radiation
G = Soil thermal flux under the canopy layer
e_a = Water vapour pressure
T_s = Ground radiant
T_a = Air temperature
e_{a*} = Air saturated water vapour pressure

γ_{ac} = Air dynamic resistance

γ_{bh} = 4/u; u is wind speed at the altitude of two meters

Δ = Slope of relation between saturated water vapour pressure and temperature

γ_c = Resistance on the canopy layer

γ = Dry-wet ball constant

ρ = Air density

C_p = Air specific heat under certain pressure

where the albedo and ground radiant temperature were derived from the NOAA-AVHRR image.

Among the five channels of NOAA-AVHRR, the values of both channel 1 and channel 2 (visible light and near-infra-red) are shifted into emissivity through calibration, and the values of channels 3, 4 and 5 change into light radiance through calibration.

Evapotranspiration and soil moisture content are calculated by the procedure presented in figure 2.

Based on our experiment, the drought severity is classified by the percentage of soil moisture divided by soil field capacity in the top 20 centimetres of soil (table 1).

Table 1. Classification of drought classes (soil moisture/field capacity, 20 cm)

Drought class	Percentage
Severe stress	<40
Moderate stress	40-50
Light stress	50-60
Normal (no stress)	60-80
Moist	>80

C. Results

By using the Crop Water Stress Index method, we monitored the drought situation of winter wheat fields in the Huang-Huai-Hai Plain Zone in 1995. Thus, the drought situation distribution map of each "decade" (ten-day period) during the growing period from February to June and information about the drought situation of various counties, cities and provinces were obtained.

Generally speaking, the percentage of drought area over planting area of winter wheat (table 2) gradually increases during the growing period, with the exception of the second and the third decades (11-20, 21-30) of April.

From table 3, we can see the winter wheat drought severities in different municipalities and provinces in HHH region. For example, in the third decade of May, the drought situation in Henan Province is very severe -- the drought area occupies 68 per cent of the arable land, while the percentage of drought area in Hebei and Anhui is 41 per cent and 34 per cent, respectively, but there were no droughts occurring in Beijing or Tianjin.

Figure 3 corresponds to the drought distribution map of Huang-Huai-Hai plain in February 1995. This distribution of different drought severities is very clear. There are boundaries of provinces and municipalities to make it easy to understand the drought situation of each province. One map shows the drought situation of a decade. Thus, with a series of maps of different periods (figures 4, 5, 6 and 7), we may see the trend and dynamics of the drought situation in the same area.

D. Conclusion and discussion

Having done field experiments for two years, we found that it is feasible to assess soil moisture by this method and that the accuracy reaches around 82 per cent. By combining NOAA-AVHRR images with ground meteorological data, the evapotranspiration ratio (E/Ep) and CWSI can be obtained.

Table 2. Percentages of drought acreage (hectare) over winter wheat sowing acreage in HHH region, 1995

Month	10-day period	Severe stress (per cent)	Moderate to stress (per cent)	Light stress (per cent)	Total area under stress (per cent)
February	1	2.59	0.29	1.43	4.22
	2	0.90	0.53	2.42	4.85
	3	0.65	1.46	5.01	7.12
March	1	1.72	2.88	7.30	11.90
	2	3.16	2.81	10.13	16.10
	3	4..03	7.14	15.47	26.64
April	1	6.04	8.72	16.61	31.37
	2	3.50	2.84	5.92	12.26
	3	1.03	3.03	15.72	19.78
May	1	3.21	7.27	18.71	29.19
	2	6.65	8.39	15.08	30.12
	3	7.72	9.79	20.37	37.88
June	1	10.43	16.91	14.14	41.48

Table 3. Percentages of drought acreage/winter wheat sowing acreage (decade 3, May 1995)

	Severe stress	Moderate stress	Light stress	Total area under stress
Beijing	0	0	0	0
Tianjin	0	0	0	0
Hebei Province	3	8.0	30.0	41.0
Jiangsu Province	0	0	13.0	13.0
Anhui Province	0	8.0	26.0	34.0
Shangdong Province	1.0	1.0	12.0	14.0
Henan Province	26.0	24.0	18.0	68.0

Through statistical analysis, the following correlation between CWSI and soil moisture content was established:

$$W = 21.3 - 15.1 \times CWSI, \quad R = -0.694, \quad N = 262 \qquad (5)$$

where W = soil moisture content.

However, field validation and more ground truth checks are needed for further improvement of this method.

The authors also recommend further research in comparative studies of different methodologies for drought assessment and monitoring.

Reference

Tian Guoliang (1990). *Dynamic Research on Remote Sensing.* Scientific Press.

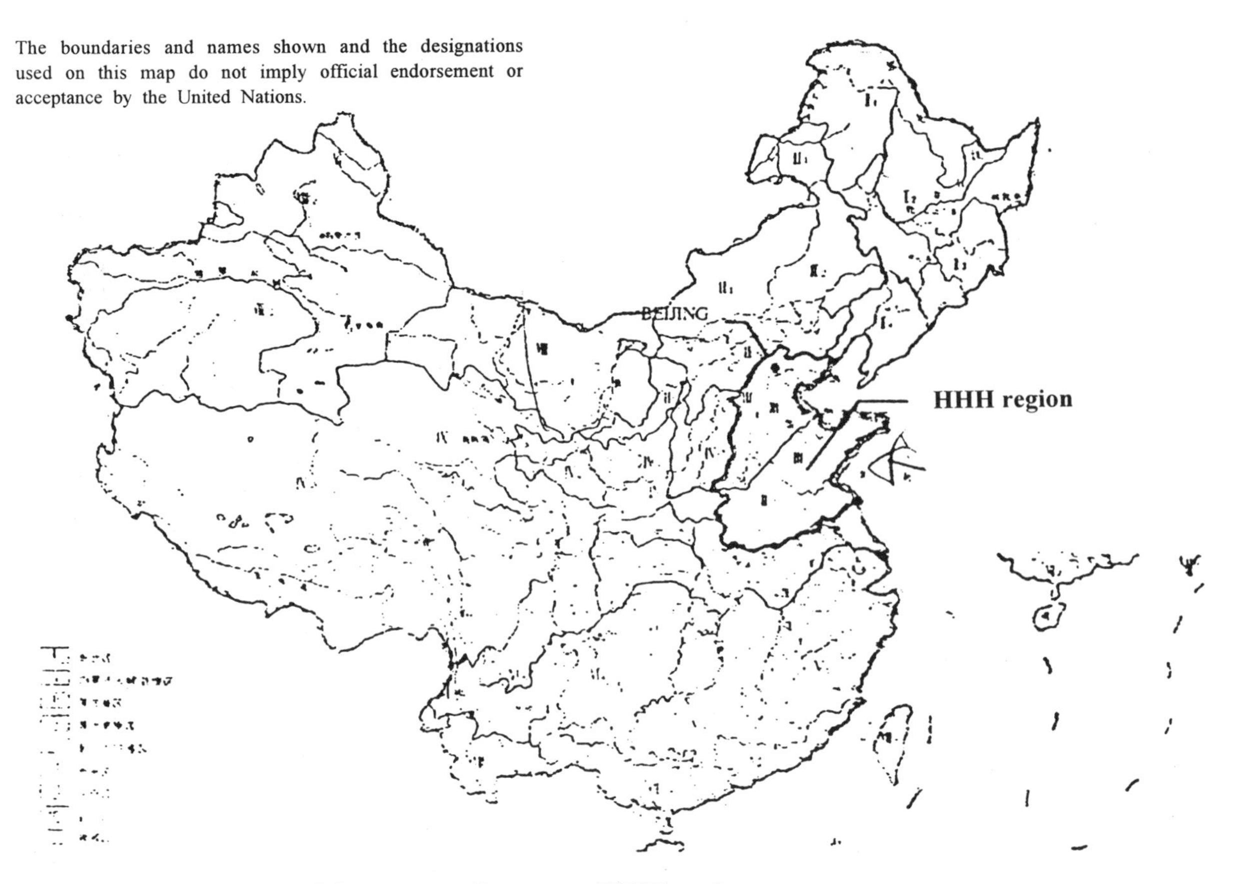

Figure 1. Location of the case study area -- HHH region

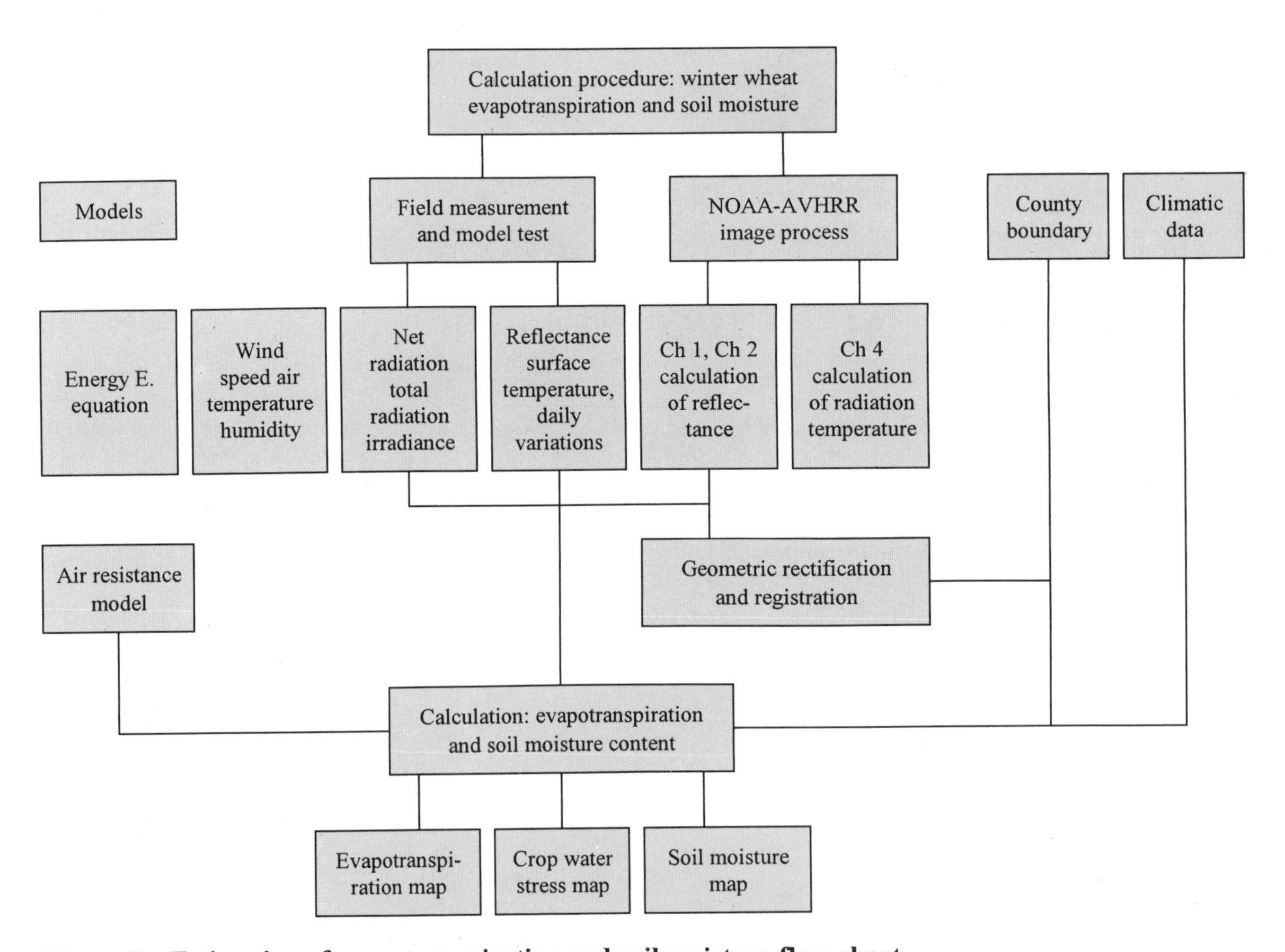

Figure 2. Estimation of evapotranspiration and soil moisture flow chart

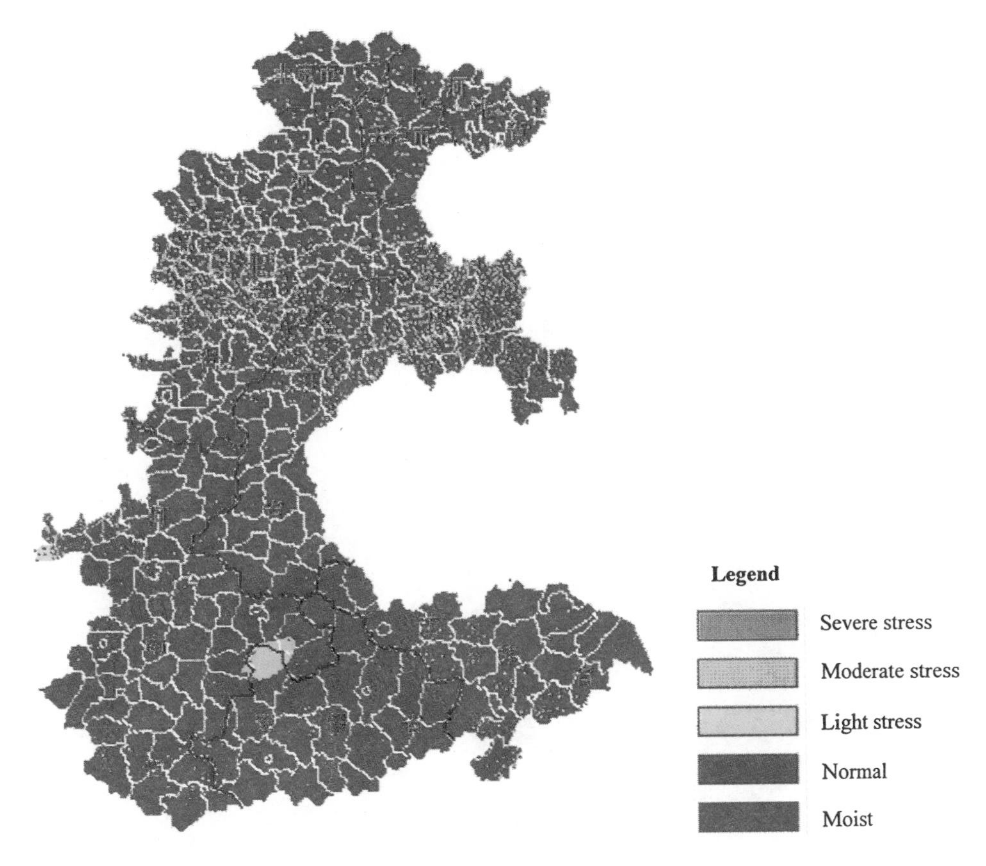

Figure 3. Drought distribution map of Huang-Huai-Hai, 10 February 1995

Figure 4. Drought distribution map of Huang-Huai-Hai, 10 March 1995

Figure 5. Drought distribution map of Huang-Huai-Hai, 10 April 1995

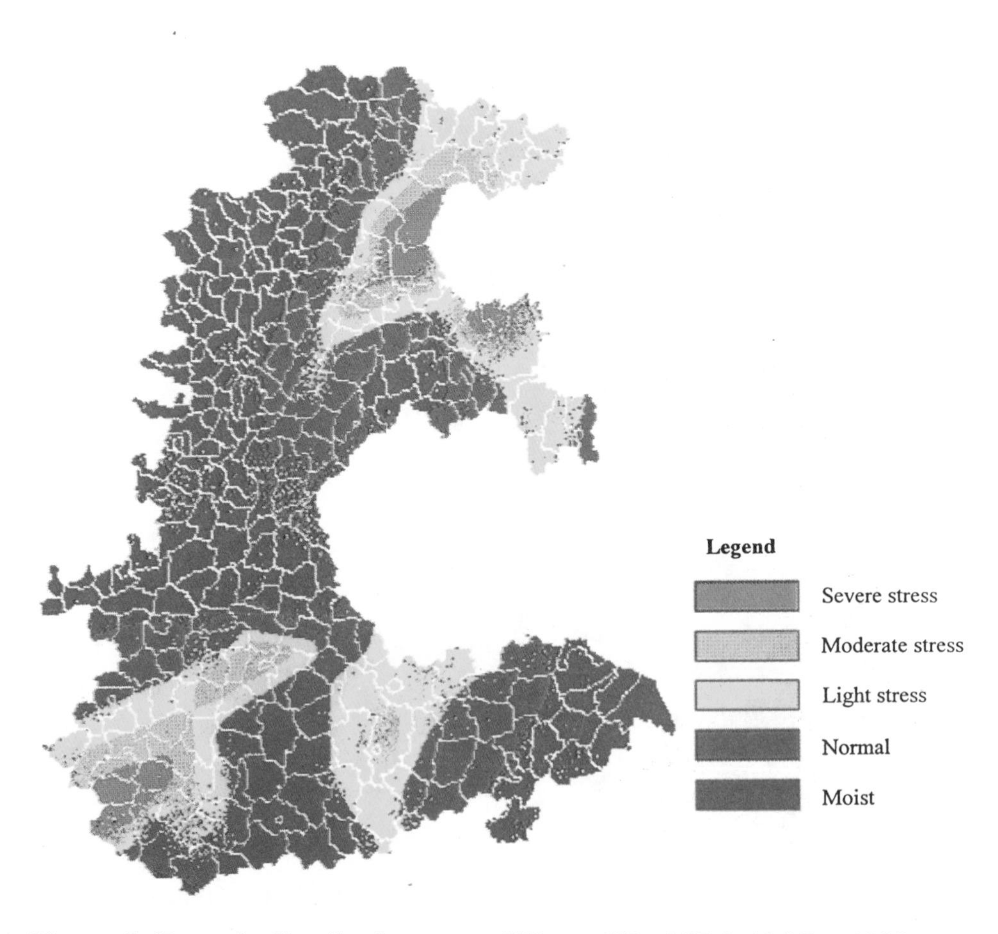

Figure 6. Drought distribution map of Huang-Huai-Hai, 10 May 1995

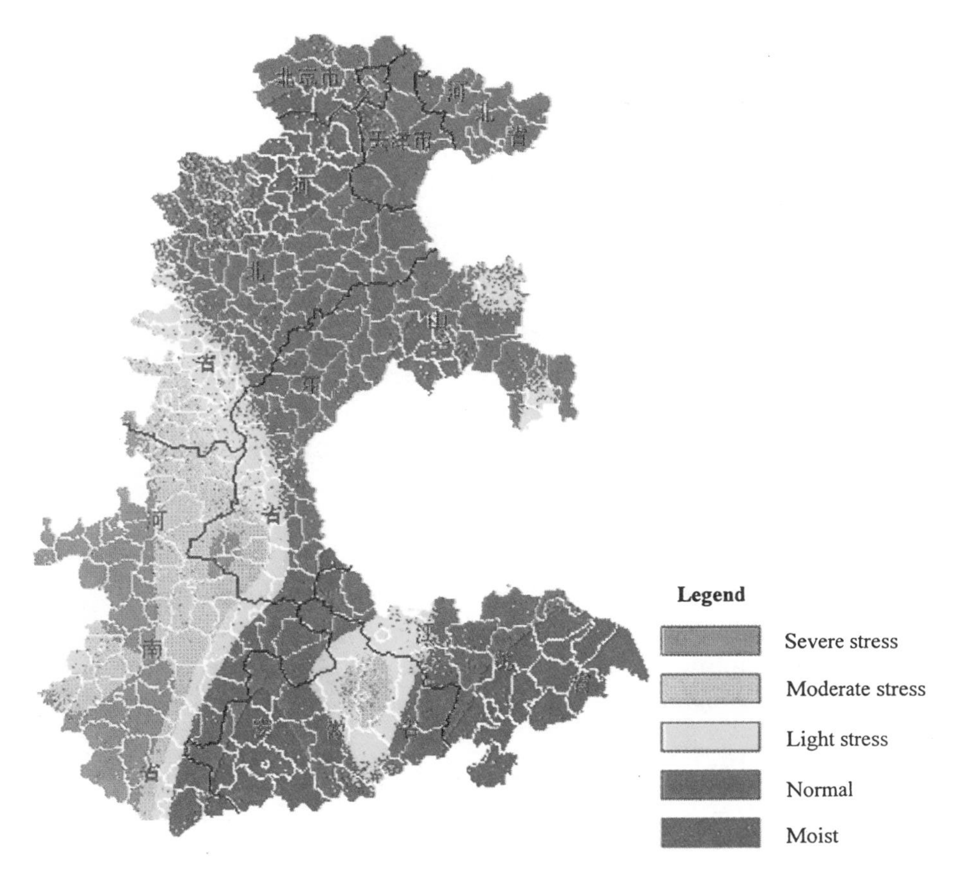

Figure 7. Drought distribution map of Huang-Huai-Hai, 10 June 1995

REMOTE SENSING AND GIS IN SOIL APPLICATIONS

*L. Venkataratnam and B.R.M. Rao**

A. Introduction

Soil is one of the important basic natural resources and information on soils is needed for a variety of purposes, because information plays a vital role in agriculture and non-agriculture programmes. In recent years, soil and environmental degradation is increasing for various reasons, which is a major concern for planners, environmentalists and scientists. Coupled with these problems, ever-growing population and increased demand for finite natural resources have aggravated the problem. To meet the challenges of the future and to maintain the productivity levels of natural resources, the emphasis for planners is on management of land/soil and water resources on a sustainable basis. Advancements in scientific and technical fields have added new dimensions to various aspects of these resources and their management. Rapid developments in space and computer technology, especially remote sensing technology, have opened up new vistas in the preparation of inventories on soil, water and biotic resources, and information so generated is being increasingly used for various development programmes.

B. Remote sensing and soil resources mapping

Soil resource information is generated basically through soil surveys. Comprehensive knowledge about soils, especially their inherent physical, chemical and morphological properties, capabilities and limitations, is needed for management of natural resources on a sustainable basis. Conventional soil surveys provide such information at a very high cost, and they take a great deal of time. Systematic application of space-borne remote sensing data in soil resource mapping enables scientists to develop operational methodologies to map soils on a routine basis and to map and monitor degraded soils like salt-affected soils, eroded soils, waterlogged areas and others. These methods become possible with inherent advantage of satellite remote sensing. Of late, geographical information system (GIS) has become a powerful tool in storing voluminous spatial data (e.g. soil maps, land-use/land-cover maps) and non-spatial attribute data (e.g. soil properties, rainfall, temperature data, socio-economic data), as well as in integrated analysis of resource data and in development of sustainable action plans.

C. Visual interpretation of satellite data

The visual method is commonly employed in interpretation of remotely sensed data from various sensors -- such as Landsat-MSS, TM, IRS and SPOT, which are based on photo-elements like tone, texture, size, shape, pattern, aspect and association -- in the preparation of soil maps at various scales, in association with published geological maps and Survey of India topographical maps. Following this approach, soil mapping at 1:1,000,000 scale is attempted by NBSS and LUP, Nagpur. NRSA has prepared nationwide soil maps of salt-affected land, using Landsat-TM/IRS data at 1:250,000 scale, in association with NBSS and LUP, Nagpur, and other state and central government organizations. In another national project, Integrated Mission for Sustainable Development (IMSD), a soil resource map is being generated at 1:50,000 scale and being used as basic data in the development of action plans for sustainable development.

D. Digital techniques for mapping soil resources

Very limited work has been reported in the literature on mapping soils using digital analysis techniques (NRSA, 1994). These studies exhibited different results. Hence, efforts are being made

* Agriculture and Soils Group, National Remote Sensing Agency, Balanagar, Hyderabad, India.

in the direction of developing digital techniques for soil mapping because of certain inherent advantages. Digital techniques allow correct radiometry, maximum spatial resolution and a high degree of maneuverability, and they facilitate discrimination of soil classes and their phases of degradation. Reliance on digital techniques is quickly increasing to cope with the enormous data inflow and increasing demands for planning.

Digital techniques rely solely on spectral response of a soil; it has been reported that the spectral signature for the same type of soil was found to vary with the change of solar elevation angles, vegetation cover and moisture conditions. The earlier studies on digital analytical techniques for soil classification revealed that soil with vegetation cover have been misclassified. To overcome difficulties in digital techniques, efforts are going on to develop context classifiers, decision tree classifiers and others, along with application of image enhancement techniques. The future generation of satellites -- with higher spatial, spectral, radiometric and temporal resolutions, advancements in image processing and analysis of microwave data -- offers greater scope in the development of methods for digital mapping of soils.

E. GIS application in soil resource studies

The land evaluation principle is based on matching the requirements of a land for specific use with the characteristics of inherent soil, climatic, topography and other natural resources and is concerned with the assessment of land performance when used for a specified purpose. Remotely sensed data is regularly used to assess the natural resources of a region and provides both spatial and attribute data with limited field work. Development of computer technology, especially GIS enables researchers to store voluminous data and enables integrated analysis of data. In the land evaluation process, GIS has become an important tool. Major GIS applications in the field of soils are land capability classification, land irrigability classification, watershed management and generation of optimal agricultural land use planning. Results of some of the case studies are discussed below.

1. Land capability classification

In one of the GIS studies at NRSA, a land capability map was generated for the Ibrahimpatnam block of Ranga Reddy District, Andhra Pradesh. The standard USDA land capability classification approach was adopted in the study. Various derivative maps covering the themes of soil erosion, slope, depth, excess water, rock outcrops and gravel stones were generated from a soil maps and other related information at 1:50,000 scale using aerial photographs. The derivative maps were prepared using the IDRISI GIS package on a PC system, and a land-capability map was generated by performing integrated analysis of the above-mentioned thematic layers. For the test site, a total of six land-capability classes were generated.

2. Watershed management

A watershed is a natural hydrological unit and is considered more rational for land and water resource development because the resources have optimum synergetic interaction. Watershed prioritization for soil and water conservation and development of an action plan for a watershed are discussed below.

Sedimentation in reservoirs, lakes and tanks at faster rates than expected is an indicator of the continuing land degradation in the catchments. Because of insufficient funds and manpower for natural resource surveys and for soil and water conservation programmes, watersheds in the catchments have been prioritized, and watersheds with high priority are treated to control soil erosion and sediment flow into the water bodies. The important parameters considered for prioritization of watersheds are physiography, slope, soil properties like texture and depth, land use/land cover, soil erodibility and existing soil conservation measures. A GIS approach has been used in the prioritization of watersheds.

In one of the recent studies at NRSA, the sub-watersheds in part of the Sileru-Machkund catchment in Visakhapatnam District of Andhra Pradesh have been prioritized through a combination of remote sensing techniques and GIS approach. A physiography map, slope map, soil map and land-use/land-cover map were generated for part of the test site using remotely sensed data of the IRS-1B L2 sensor along with published geological maps and Survey of India topographical maps. Subsequently, various thematic map, spatial data and attribute data generated were used in the creation of the database in GIS. Arc Info software package version 5.0/6.0 was used to digitize, edit and display for integrated analysis and to generate output plots. Watersheds were prioritized following the Sediment Yield Index (SYI) approach (AIS and LUS). SYI is given by:

$$SYI = \frac{(EI \times A_{ie} \times D)}{AW} \times 100 \qquad (1)$$

where

SYI = Sediment Yield Index
EI = Weightage value of erosion intensity unit
A_{ie} = Area of erosion intensity/composite unit
D = Delivery ratio
AW = Total area of the watershed.

These composite erosion intensity units were assigned weightage values following the standard approach and were added to the attribute table. A delivery ratio was given to each composite unit depending on its nearness to a reservoir and was incorporated in the attribute data table. The erosion intensity map generated was combined with a watershed map and a new map was generated with watershed areas as attributes in addition to other attributes. The calculation operation was used to compute SYI value for each watershed, using the above formula. Watersheds with SYI values were classified following the standard AIS and LUS approach. Out of eight sub-watersheds in the watershed, four watersheds are rated "very high priority", three are rated "high priority" and one is rated "medium".

F. Integrated management of agricultural resources

To manage land and water resources for suitable agricultural productivity in a rational manner, by adoption of appropriate strategies and technologies, requires an understanding of the mutual interdependencies of renewable and non-renewable resources, integration of the land and water resources and identiflcation of the constraints and ecological problems at the micro-level. The locale-specific solutions can be arrived at through the effective use of space-based remote sensing data combined with other socio-economic data using geographic information systems. This is being done by a survey of resources at 1:50,000 scale using traditional and remote sensing techniques, evaluation of collected data, preparation of a set of resource maps (hydro-geomorphology, soils, land use and cover, surface water, drainage, watershed and other types) and generation of an action-plan map giving site-specific recommendations for development of agriculture, groundwater recharge, fuel and fodder, soil conservation and afforestation. It is through implementation of such site-specific reclamation/conservation measures and management practices that productivity at the micro-level can be sustained on a long-term basis.

Recently at NRSA (1996), a sustainable action plan for part of the upper Machkund watershed, falling in tribal areas of the Visakhapatnam District, Andhra Pradesh, was generated through GIS techniques. For the sustainable development plan, four action items, i.e. afforestation, soil conservation plus silvipasture, soil conservation plus agro-horticulture, and double cropping, were selected. Rules have been framed to combine geomorphology maps, groundwater potential maps, soil maps, land-use maps, drainage maps and others in an Arc Info system. Various derivative maps, namely soil-texture map, land-capability map, soil-reaction map and depth map, were generated. The action plan for the watershed revealed that 54 per cent of the watershed needs afforestation, 11.4 per cent soil conservation and silvipasture, 1 per cent soil conservation and agro-horticulture, and 24 per cent double cropping.

PART SIX

FORESTRY AND LAND DEGRADATION

APPLICATION OF REMOTE SENSING AND GEOGRAPHIC INFORMATION SYSTEM IN FOREST STUDIES

*P.S. Roy**

A. Introduction

Forests and woodlands cover nearly one-third of the Earth's total land surface. The fuel, food and income they provide are basic to the well-being, even the very survival, of hundreds of millions of people. Yet, of all the world's natural resources, the forests and woodlands are perhaps the most neglected and are being depleted at an alarming rate.

The phenomenal growth of human population and the number of livestock in recent decades has placed the forests of India under a pressure that is beyond their carrying capacity. Degradation and deforestation threaten environmental stability and ecological security.

The spatial distribution of forests in India is uneven. After independence, the country suffered rapid deforestation, especially because of increases in agriculture production. The remaining forest lands have suffered continuous depletion because of the staggering gap between demand and production of firewood. Grazing pressure due to the more than 400 million cattle in the country has also caused serious damage to species regeneration and forest quality.

1. Priorities for forest resource management

For effective and systematic forest resource management, a strong database is necessary, which can be easily stored, retrieved, analysed and accessed. Presently, the requisite information for such a database is not available. The following are the priorities for the effective management of forest resources:

(a) A tool for constant and regular monitoring of the forest's status;
(b) A habitat management system for protected areas focusing on total biodiversity maintenance;
(c) An effective afforestation planning for unproductive (degraded) lands;
(d) A management strategy for grassland ecosystem evaluation;
(e) Evolution of a fire risk-model to map fire-prone areas.

B. Role of remote sensing

A better knowledge of forests implies information about their potential, extension, composition and evolution, including their rate of transformation to other uses. There is a need to obtain reliable data about vegetation resources at regional and micro-levels, which would help in planning a forest management strategy for sustained yield so as to benefit society. Satellite remote sensing is a timely technological development, in view of the serious pressure on the nation's natural resources. In order to meet the increased need for application of remotely sensed data in India, the Department of Space plans to have indigenous remote sensing platforms capable of acquiring data in visible, near-infra-red, middle infra-red and microwave regions of the electromagnetic spectrum. IRS-1A and 1B are state-of-the-art satellites. IRS-1C is a second-generation operational remote sensing satellite with improved features, such as better resolutions, sensing capabilities in middle infra-red, stereo-viewing, revisit capabilities and onboard data recording. India also has plans to build a full-fledged microwave remote sensing satellite after sufficient research in the working of sensors, their data processing and interpretation aspects for applications.

* Indian Institute of Remote Sensing, National Remote Sensing Agency, Dehra Dun.

Generally, current forest maps have been found too lacking in accuracy, appropriate details and timeliness to be used effectively for forest management requirements (Houghton and Woodwell, 1981). Recent developments in remote sensing technology have indicated that if it is judiciously combined with ground-based studies, it is possible to carry out detailed forest inventories and monitoring of natural vegetation cover at various scales (Botkin et al., 1984) (table 1).

Table 1. Remote sensing approach for managing forest resources at different levels

Information level	Aspects	Data source	
		Ground-based	Remote sensing
Global	– Biogeographic classification – Vegetation dynamics – Climatic influences – Overused areas	Less	Satellite
Regional	– Forest cover mapping – Vegetation type mapping – Habitat analysis – Pattern diversity – Land use – Monitoring	Moderate	Satellite
Local	– Resource inventory – Hotspot monitoring – Network of protected areas – Corridor identification – Site suitability – Species information base	High	Satellite/aerial

Looking back at the problems related to deforestation, global climate change and regional/micro-level forest management requirements, we shall adduce studies using remotely sensed data from various satellite platforms that cater for the needs of Indian forest managers and ecologists.

1. Macro-level vegetation cover monitoring

Studies indicate the possibility of estimating the global vegetation cover based on meteorological satellite data. The polar orbiting satellite NOAA-AVHRR (the National Oceanic and Atmospheric Administration's Advanced Very High Resolution Radiometer) provides digital data in the visible, near-infra-red and thermal channels of the electromagnetic spectrum. The normalized difference vegetation index (NDVI) is one commonly used ratio electromagnetic data, which is highly correlated with vegetation parameters such as green leaf biomass and green leaf area and hence is of considerable value for vegetation discrimination (Holben et al., 1980):

$$NDVI - (CH_2 - CH_1)/(CH_2 + CH_1)$$

The normalized difference vegetation index has been used in India to identify forest cover (Roy and Kumar, 1986). The forest-cover maps generated by visual interpretation of Landsat MSS false colour composites (FCC) and NOAA-AVHRR NDVI show close similarity between closed forest, open/degraded forest and non-forest areas. A technique for preparing potential vegetation-cover maps using NOAA-AVHRR (LAC data) -- the multi-date maximum vegetation index -- has been developed in India. These maps can be of immense value of studying vegetation, its phenology and biomass.

2. Regional applications of remote sensing

Satellite remote sensing has played a key role in providing information about forest cover and vegetation type, and their changes on a regional scale. Initial studies indicated the absence of an appropriate classification scheme. Subsequent developments in image processing techniques, understanding of the temporal characteristics of responses and standardization of ground sampling methods have brought profound acceptance of the application of remote sensing data in forest inventorying and mapping at the regional scale. The immediate forest management requirement is to create a database for available forest resources in spatial terms. It would help in defining forest policy in a scientific manner.

(a) Forest cover mapping and monitoring

Landsat MSS data have been used extensively to delineate forest and non-forest areas in the tropical countries where up-to-date data regarding spatial distribution are lacking or inaccurate. The studies used manual interpretation of individual images at a scale of 1:1,000,000 or 1:250,000. Such a study was carried out in India by the National Remote Sensing Agency in 1983, the first of its kind, to show the potential of remote sensing technology at the national level. The study mapped natural tree areas into three categories -- closed forest, open/degraded forests and mangrove forests -- using Landsat MSS false colour composites of the periods 1972-1975 and 1981-1983 (table 2).

Table 2. Forest cover of India estimated using satellite remote sensing data

Forest category	Forest area in terms of percentage of total geographic area	
	NRSA estimates 1972-1975	**FSI estimates 1981-1983**
1. Dense forest (per cent)	14.12	10.88
2. Open forest (per cent)	2.67	8.41
3. Mangrove forest (per cent)	0.99	0.12
4. Coffee plantation (per cent)	--	0.11
Total forest (per cent)	17.78*	19.52

Note: * Scrub and coffee plantation not included.

Subsequently, the Forest Survey of India (FSI), an operational agency set up to execute such tasks, used a similar data set to map the forest cover of India in 1980-1983. Interestingly, both mapping exercises showed almost the same figures for closed forests and mangrove forests.

The study thus proved that the technology can be used for monitoring the forest resources of a vast country such as India. The Forest Survey of India has now formulated a programme to monitor the nation's vegetation cover and to develop a database that will use satellite remote sensing as one of the prime data sources.

(b) Vegetation type mapping

Forest vegetation types have been described on the basis of physiognomy, structure, function and composition (Forsberg, 1967). Vegetation has also been classified based on height classes and distinction of woody tissue (UNESCO, 1973). The vegetation types of India have been classified by Champion and Seth (1968) based on physiognomy, following climate, succession and ecological status. Adoption of a classification scheme in remote sensing for vegetation mapping will require simplification of the above approaches.

(c) Manual interpretation

Vegetation type mapping using visual interpretation of satellite data can be adjusted to the requirements or objectives of the survey. It is possible to conduct condition mapping by highlighting the structural aspects of the forest stands or to conduct type mapping by highlighting the floristic composition formations of the forest stands.

(d) Enhancements

There are well-defined image enhancement techniques that greatly increase the amount of information that can be visually interpreted from the image data by improving the apparent contrast between features in the scene. The enhanced images are also of immense use in improving ground sampling design. Important enhancement techniques that have found application in forest vegetation type mapping are the following:

(a) Contrast stretching increases the separation between two vegetation type boundaries allowing higher delineating accuracy;

(b) Hue-saturation-intensity (HSI) transformation of a normal data set increases the colour differentiation between different vegetation categories;

(c) Principal component (PC) transformation compresses the multispectral response and provides more differentiation between the vegetation types;

(d) Kauth-Thomas transformation shows pronounced differentiation between vegetation types. The separability was confirmed from bivariate plots of true ground observations (Roy et al., 1991a). However, the derivatives were found to be sensitive to shadow patterns.

(e) Digital classification

Visual interpretation, although it has certain advantages, may not be used for regular monitoring on an operational basis because its inherent subjectivity is a drawback. Digital techniques largely eliminate this drawback and are more useful in monitoring.

Digital handling of data enables accurate processing of multispectral data, because plant species produce distinguishable spectral reflectance differences in one region of the electromagnetic spectrum but not in another. The Landsat MSS era has seen the utilization of one near-infra-red and three visible bands for broad vegetation type mapping using digital methods. However, second-generation satellites such as Landsat TM have greatly helped in stratifying forests accurately thanks to the presence of the middle infra-red channel. Roy (1987) observed that in using the normalized vegetation index along with normal multispectral data, a larger number of forest classes could be identified by digital classification. Furthermore, digital classification increased the accuracy of common classes in general. Roy et al. (1991a) used Landsat TM digital data in a comprehensive study to stratify forest types in the Baratang forest division of the Andaman-Nicobar island group. A classification using the same ground truth samples was undertaken on three data sets, namely bands 3, 4 and 5; bands 2, 3 and 4; and bands 2, 3, 4 and 5. Computerized stratified random sampling was carried out to extract field sampling points to be checked for classification accuracy evaluation.

The classes mapped and ground sample "truths" were compared to estimate classification accuracy, as detailed below:

(a) Maps from bands 3, 4 and 5 = 90.30 per cent;
(b) Maps from bands 2, 3 and 4 = 71.70 per cent;
(c) Maps from bands 2, 3, 4 and 5 = 90.90 per cent.

The accuracy figures indicate that the vegetation maps generated by digital classification of bands 3, 4 and 5 are highly reliable.

Indian Remote Sensing (IRS) satellite data were used to analyse forest types, vegetation density and land use practices in Aglar watershed, of the Garhwal Himalaya (Roy et al., 1990). Supervised maximum likelihood classification using IRS LISS-II data provided stratification of forest types and level II land use classification. The vegetation crown closure density was stratified using combined IRS-1A LISS-II four-band data and a normalized vegetation band. The study showed that the necessary information of the watershed with respect to forest type and crown closure could be reliably obtained from IRS data with an overall signature accuracy of 92.40 per cent and 90.50 per cent, respectively. It was also observed that a normalized vegetation index, if combined with a normal data set, increased the accuracy in crown closure/physiognomic stratification. The normalized vegetation index has shown a positive relationship with the crown closure and no relationship with the forest type.

Investigations of the use of synthetic aperture radar (SAR) data for qualitative forest stratification have been reported (Wu, 1981; Werle, 1989; and Muller et al., 1985), but such investigations with Indian forests are lacking. Roy et al. (1994d) tested the digital preprocessing techniques, and visual mapping capabilities of airborne X-band SAR data for diverse vegetation types in a tropical wet climate.

Spatial textural analysis methods have also been evaluated to enhance discriminability of the forest types and features. Attempts have also been made to enhance the information by merging optical remotely sensed data with microwave X-band responses. The backscattering digital values have been correlated with qualitative and quantitative vegetation parameters. A significant positive correlation with leaf area index (LAI) has been observed. The results obtained in the present study indicate that microwave remote sensing data can provide valuable information about the vegetation canopy characteristics. The microwave data from satellite platforms hold promise for vegetation mapping and monitoring and probably can address the present limitations of optical remote sensing.

(f) Analysis of forest disturbance

Tropical forests are facing disturbances of varying magnitude in different regions. Because of over-extraction of resources, tropical wet evergreen forests are losing their original structure and facing retrogression. Spectral differences in the spectral ellipse plot of bands 5 and 4 have been observed by Roy (1989) in virgin evergreen forests of Baratang Island between 1972 and 1982.

Shifting cultivation is also bringing about serious disturbances in primary forests of the north-east region of India. Roy et al. (1985) observed that it is possible to stratify primary and secondary forests along with shifting cultivation and abandoned shifting cultivation areas.

Roy et al. (1993a) made primary and secondary analyses of vegetation using remote sensing and phyto-sociological ground data collected from sample plots to assess the ecological importance of different species in the Bakultala Range of the Andaman Islands. Interrelationships between different communities have been evaluated through various available indices. The study highlights the fact that retrogression has set in at the community level and stresses the need to conserve the germ plasm present in the natural evergreen vegetation. Shirish (1994) has used satellite data for community structure analysis in Madhav National Park, Madhra Pradesh. Community structure included detailed descriptions of the floristic composition of forest types and their structural parameters (density, basal area, abundance). Quantification of these parameters ultimately contributed to a value index of species. The value index was further used for estimation of community measures (diversity index, species dominance, degree of evenness).

(g) Monitoring vegetation changes

Manual interpretation and change delineation by superimposing two or more time period maps have been carried out by remote sensing scientists since such data became available. A study in forest land use change carried out in Baratang forest division of Andaman and Nicobar islands highlights the land transformation and its influences (Roy et al., 1991a). The results highlight the human role in altering the land cover during the last 18 years. Extraction of commercial forest resources is leading to change in types, conversion of forest land to agriculture land use, and conversion of forest areas to monospecies forest plantations. The extraction of virgin evergreen forest has led to a retrogressive successional trend in the vegetation.

2. Micro-level applications of remote sensing

(a) Stock mapping

Indian forests should continue to be managed through working plans that are prepared or revised at 10-15 year intervals with a grassroots level information base of stock maps at a 1:15,840 scale (4" = 1 mile). The stock maps depict forest types, density encroachment, cultivation patches, regeneration status and some idea of available resources. The efforts involved in the preparation of stock maps through conventional ground survey methods are time-consuming and strenuous.

The boundaries of forest types, drainage, encroachments and cultivation patches are much more reliable. IRS LISS-II data have been used to provide forest stock information in individual compartments by digitizing compartment boundaries on geometrically corrected digital data. Further processing through supervised classification and vegetation indexing of IRS LISS-II data overlaid with compartment boundaries makes such products extremely useful for practical field forestry. A recent analysis of survey requirements in forestry indicated that the stock map requirements may be accomplished with a 1:25,000 scale (Porwal and Roy, 1995).

(b) Growing stock estimation

In India, there has been a shift of priority from commercial forestry to conservation forestry in the last decade. This has resulted in a low priority for resource inventories in the form of growing stock estimation. However, resource inventories should continue to be required in resource-rich areas that can provide timber for commercial purposes. Ground-based inventories have given way to inventories through aerial photographs using stratified sampling. Forest resource stratification using satellite remote sensing data have proven invaluable in carrying out multiphase sampling. Stratified random sampling based on a visually interpreted map from Landsat TM FCC at 1:50,000 scale was used by Singh and Roy (1990) in the south Andaman forest division of the Andaman and Nicobar islands. Group sampling in 0.1 ha sample plots was carried out to estimate the volume of individual tree species and subsequent grouping in commercial and non-commercial species. The sample volume data were enlarged to stratified population on remote sensing data. The growing stock estimations were found comparable to the Forest Survey of India estimates (1981).

(c) Biomass estimation

India's rural population is dependent on fuelwood for their daily energy needs. It is estimated that about 150 million tons/year of fuelwood is required for the present population. According to estimates, about 80 million tons/year is met through forests, private holdings, agricultural wastes and animal wastes, and over 60 million tons/year of the total requirement is met through kerosene, petroleum and coal. The remaining shortfall in the fuelwood requirement is met by illegal harvesting of biomass from the forests. It is felt that biomass consumption for meeting the energy needs of rural India is the most practical and inevitable. For this purpose, inventories are essential, which can provide baseline data about available vegetation resources. Biomass distribution is primarily a function of biomass, vegetation type, climate and site conditions. As a result, initial stratification available on remote sensing data can be combined with ground sampling. Roy et al. (1994b) attempted to overcome limitations of conventional methods of biomass estimation. Two different approaches were developed:

(a) Statistical sampling approach;
(b) Spectral response-based model.

Regional biomass maps have been produced by both techniques. Multiple regression models have been developed using greenness, brightness and wetness components isolated from Landsat TM data. In both approaches, methods to find homogeneous vegetation strata (HVS) wherein ground sampling should be done has been suggested. It can be said that rapid biomass inventories can be made using 1:50,000-scale satellite images with minimum non-destructive field sampling. For this purpose, one basic requirement is the availability of generalized species equations for dominant species and generalized interspecies equations for subordinate species.

(d) Habitat analysis

The essential parameters for habitat analysis are cover, food, water and topography. Some of these parameters have been collected from remotely sensed data in Kanha National Park and Rajaji National Park to classify the areas in various habitat zones. Besides information acquired from remote sensing data, field evidence of animal presence and biotic disturbance have been used for habitat analysis. Based on specific needs of wild animals, the grid-based criteria indexing method has also been adopted to derive a habitat suitability index in Kanha National Park for wild animals such as the gaur, sambar, bison and tiger. In a recent study by Roy et al. (1994c), a geographic information system was used in studying suitable habitats of goral (mountain goats) in Rajaji National Park (Uttar Pradesh). Slope/terrain, food and water availability, interspersion and juxta-position of vegetation types were used to model the suitable habitats for goral.

More than half of India's wetlands have been lost and more than half of the mangroves have been cleared or degraded, many for conversion to aquaculture ponds. As a result, wetlands are probably the most endangered habitats in the tropics. The Indian Institute of Remote Sensing and the Salim Ali Centre for Ornithology and Natural History have carried out a study in Etawah and Mainpuri districts for estimating the conservation value of wetlands (Das et al., 1994). The major goal was to investigate whether the combination of remote sensing and ground data would enable classification of wetlands on an ecological basis. The focus was on mapping of wetlands, vegetation/ land use and fauna (fishes and birds). Based on a GIS database of wetlands spatial characteristics, the surrounding land use, bird population diversity and evenness and plant community distributed in 18 wetlands were identified in two districts for conservation.

(e) Landscape analysis

Ecologist have given prime importance to pattern diversity so as to maintain total biodiversity. Remote sensing provides information with respect to homogeneous landscape units which can be evaluated in a spatial context through landscape analysis. Remote-sensing-derived vegetation maps provide a perspective horizontal view and help in delineating different landscape elements and their characteristics. Shirish and Roy (1993) have used satellite-derived vegetation and physiographic maps for analysing landscape elements of Madhav National Park (Madhya Pradesh) using a GIS. Patch characteristics (shape, size, porosity and patch density) were used to assess the disturbance gradient. The patchiness and shape of different vegetation types also provide information about the climax, serial and retrogressive vegetation forms.

(f) Forest fire prone area mapping

After deforestation, forest fires are the next most important cause of incalculable harm to extensive forest areas. The annual recurrence of forest fires often causes irreversible damage to the environment, endangering regeneration and at times causing a total loss of vegetal cover. The burning associated with fires can lead to higher concentrations of greenhouse gases, accounting for roughly one quarter of all CO_2 build-up (Bolin, 1986). The main factors influencing fire risk are vegetation cover, human interference and topography. Moreover, the fire's behaviour, its rate of spread, direction of travel and intensity largely depend on fire environment, that is, weather, insula-tion and wind. A study made by Abhineet et al. (1996) was aimed at investigating a reliable model for fire prone area mapping in the Dhaulkhand range of the proposed Rajaji National Park (Uttar Pradesh). The approach involved the integration of remote sensing and field survey data into a geographic information system in order to find the relation between various factors influencing fires. The model used and the resulting fire prone areas map will be of immense assistance to the fire fighting authorities in formulating remedial measures and will certainly enable the planners to draw up more efficient protection programmes.

(g) Evaluation of grassland bioproductivity and carrying capacity

Range inventory and monitoring are essential to provide direction for eco-management of grasslands. Satellite data have been used to map grasslands and to evaluate their ecological status under the Remote Sensing Application Mission of the Department of Space. Four such studies have been carried out in the western Ghats and in various arid, semi-arid and alpine zones. Visual and digital interpretation helped in mapping degradation due to erosion and overgrazing. In the alpine pastures of Kinnaur District, the technique has been found useful in stratifying the grasslands into low-level pastures and pastures dominated by grass or shrubs or mixed. Grassland type stratification in combination with a land-form map has been used for determining ground sampling points for biomass estimation. Biomass estimation provides the productive potential of these grasslands, which in turn has been used for carrying capacity estimation (Roy et al., 1994a).

Direction of change in condition of grasslands can be related to stages in succession and retrogression of vegetation in response to management practices. Such changes can be studied by monitoring rangelands in different seasons of the year and in the same seasons of different years, and changes can be assessed to describe whether a rangeland is climax or potential natural vegetation. Replacement of grasslands by *Shorea robusta, Butea monosperma* and *Lagerstromia parviflora* in the grasslands of Kanha National Park has been detected using satellite data, which indicate that protection of the habitat is allowing climax vegetation to regenerate and that the area under grasslands is being reduced by herbivores (Roy et al., 1991b).

(h) Sustainable development

Planning of development activities has taken place through a sectoral approach. This has resulted in serious environmental problems in both rural and urban areas. In the process, the country has also extracted renewable natural resource beyond their productive potential. There is complete agreement that if development is to be sustained for a longer period, land use planning has to be evolved through an integrated approach. Such an approach will account for biological, physical and socio-economic factors.

The Department of Space, under its national programme, has undertaken the Integrated Mission for Sustainable Development (IMSD) in selected districts so as to meet critical development constraints. Wastelands are priority areas for plantations and for generating biomass resources in rural environments. For their development, it is necessary to have the proper mixture of vegetation, sustainable on a particular site, with simultaneous attention to the water regime, soil conservation, animal husbandry and the efficient use of fuel, alternative energy and fodder resources.

The planting of species should be planned in such a manner that native plant species diversity is maintained.

Wasteland development with an integrated approach requires basic data about various components of the system. Ironically, such data were rare until the emergence of remote sensing technology, which hampered the design of wasteland development programmes. Remotely sensed data at an appropriate scale promise to generate most of the required spatial data and are also amenable to compartmentalization for developing information systems for long-term monitoring. Detailed data are required on aspects such as physiography, topography, soil type, soil erosion status, land cover, land-use pattern, land tenure system, hydrology, climate, human population, cattle population, socio-economic factors, infrastructure, demand and supply, crop type, past and present growth/yield, productivity and inventory of species suitable for plantations in a particular locality (Roy et al., 1993a and 1993b). Such a database is required for multilevel planning, i.e. block, district, state and national levels in small synoptic scale or in large scale for small areas. GIS has a significant role to play in the development and planning programme. Ground-based data and socio-economic data can be analysed in an integrated manner in a GIS for developing environmental and economic strategies. The GIS data sets of the required environment can be visually or numerically displayed for exloring relationships among spatial data sets, so as to identify locations that meet specific criteria, for trade-off decisions and for assessing the impact of a proposed project or action. Such an approach will lead to sustainable development of the region and put a check on the erosion of biodiversity.

C. Conclusion

Rapid depletion of forest resources and their biodiversity in the tropics has necessitated improved land use planning and land use classification so that forest resources can be used for the benefit of mankind. Remote sensing technology will play a key role in surveying, assessing the resources and in recording the changes. Such data will also help in developing biodiversity management (figure 1). Data from geographic information systems and ground measurements, in combination with remotely sensed data, can provide information over large areas about environmental changes, while models can make future predictions about the environment.

The present state of the art can meet most of the forest management requirements of various levels. Micro-level management needs, however, are only partially met, owing to the non-availability of high resolution multispectral data. IRS-1C data, with improved spatial resolution in multispectral and panchromatic bands, will play an important role in satisfying these micro-level management needs. Its applications in real problem solving for forest management are yet to take off. The following strategies need to be followed to manage forest resources and their biodiversity:

(a) Subdivide land into manageable and meaningful regions with a variate of life-zone systems. This would help in proper forest land use planning;

(b) Create databases at various levels (macro-level, regional and micro-level) in a suitable format that can be accessed;

(c) Within each biogeographic region, select and survey pilot test sites for updated and more detailed inventories. These sites could be sites representing the entire region (using random sampling approach), sites where rapid change is occurring (e.g. deforestation) or sites that are exceptionally productive;

(d) Perform detailed ecosystem analysis so that environmental indices can be developed.

If we fail to achieve this much-needed coordinated approach, we may find ourselves at a point of no return.

References

Abhineet, Jain, S.A. Ravan, R.K. Singh, K.K. Das and P.S. Roy (1996). Forest fire risk modelling using remote sensing and geographic information system. *Current Science*, 70(10): 928-933.

Bolin, B. (1986). *The Greenhouse Effect, Climate Change and Ecosystem.* Scope 29.

Botkin, D.B., J.E. Estes, R.M. McDonald and M.V. Wilson (1984). Studying the Earth's vegetation from space. *Bioscience*, 34: 508.

Champion, H.G., and S.K. Seth (1968). *A Revised Survey of Forest Types of India.* New Delhi: Government Publications.

Das, K.K., S.N. Prasad and P.S. Roy (1994). Mapping of potential crane habitat in Etawah and Mainpuri Districts using satellite remote sensing techniques. Indian Institute of Remote Sensing report, pp. 1-37.

Forest Survey of India (1981). Report of forest resources of south and middle Andaman. FSI technical report 28, Government of India, New Delhi.

Forsberg, R.F. (1967). A classification of vegetation for general purpose. In G.F. Peterken, ed., *Guide to Check Sheet for IBP Areas,* IBP Handbook No. 473. Oxford: Blackwell Scientific Publications.

Holben, B.N., C.J. Tucker and C.J. Fan (1980). Spectral assessment of leaf area and leaf biomass. *Photogrammetric Engineering of Remote Sensing*, 46: 651-656.

Houghton, R.A., and G.M. Woodwell (1981). Biotic contributions to the global carbon cycle: the role of remote sensing. In *Proceedings* of the Seventh International Symposium of Machine Processing of Remotely Sensed Data, West Lafayette, Indiana, pp. 593-602.

Muller, P.W., R.M. Hoffer and D.F. Lozano-Garia (1985). Interpretation of forest cover on microwave and optical satellite images. In *Proceedings* of PECORA 10 -- Remote Sensing in Forest and Range Resource Management, Colorado State University, Fort Collins, Colorado, 20-22 August, pp. 578-592.

Porwal, M.C., and P.S. Roy (1995). Revision and updating of stock map using remote sensing and geographic information system. In *Proceedings* of ISRS Silver Jubilee Symposium, Dehradun, 22-24 February, pp. 334-342.

Roy, P.S., R.N. Kaul, M.R.S. Roy and S.S. Garbyal (1985). Forest type stratification and delineation and shifting cultivation areas in the eastern part of Arunachal Pradesh using Landsat MSS data. *International Journal of Remote Sensing*, 6: 411-418.

Roy, P.S., and S. Kumar (1986). Advanced Very High Resolution Radiometer (AVHRR) satellite data for vegetation monitoring. In *Proceedings* of the International Seminar on Photogrammetry and Remote Sensing for the Developing Countries, New Delhi, pp. 31-34.

Roy, P.S. (1987). Montane vegetation stratification through digital processing of Landsat MSS data. *Geocarto International* (1): 19-26.

Roy, P.S. (1989). Analysis of forest types and monitoring disturbances using Thematic Mapper data in part of Andaman and Nicobar islands. In *Proceedings* of National Seminar on Status of Indian Forestry Problems and Perspectives, Haryana Agriculture University, Hissar, pp. 22-28.

Roy, P.S., P.G. Diwakar, T.P.S. Vohra and S.K. Bhan (1990). Forest resources management using Indian Remote Sensing satellite data. *Asian-Pacific Remote Sensing Journal*, 3: 11-22.

Roy, P.S., B.K. Ranganath, P.G. Diwakar, T.P.S. Vohra, S.K. Bhan, I.J. Singh and V.C. Pandian (1991a). Tropical forest type mapping and monitoring using remote sensing. *International Journal of Remote Sensing*, 12: 2,205-2,225.

Roy, P.S., S. Jonna and D.N. Pant (1991b). Evaluation of grasslands and spectral reflectance relationship to its biomass in Kanha National Park (Madhya Pradesh), India. *Geocarto International* (1): 39-45.

Roy, P.S., S. Singh, I.J. Singh, Shefali Agarwal, L.M. Pande, K.V. Ravindran, Anita Rawat, A.K. Tiwari, S.K. Saha, P.N. Haridas, T.P.S. Vohra and K.S. Bisht (1993a). Geographic information system for wasteland development planning in Puruliya District, West Bengal: pilot project for Arsa block. Indian Institute of Remote Sensing project report, pp. 1-73.

Roy, P.S., S. Singh, L.M. Pande, K.V. Ravindran, I.J. Singh, Anita Rawat, Shefali Agarwal, A.K. Tiwari, S.K. Saha, P.N. Haridas and T.P.S. Vohra (1993b). Geographic information system for wasteland development planning in Puruliya District, West Bengal: pilot project for Manbazar II block. Indian Institute of Remote Sensing project report, pp. 1-86.

Roy, P.S., Jain Abhineet, A.R. Shirish, K.K. Das and M.C. Porwal (1994a). An approach to evaluation of bioproductivity of grasslands using remote sensing and geographic information system. Indian Institute of Remote Sensing technical report, pp. 1-33.

Roy, P.S., A.R. Shirish, I.J. Singh and S. Singh (1994b). Approach for terrestrial biomass estimation using satellite remote sensing monoculture plantations in Tarai region of Uttar Pradesh: scientific report. Global Change Studies. Indian Space Research Organization, Department of Space, Bangalore. ISRO-4BP-SR-42-94, pp. 125-148.

Roy, P.S., N.A. Rajadnya and K.K. Das (1994c). Forest type mapping, monitoring and habitat suitability analysis of goral (*Nemorhaedus goral*) using aerospace remote sensing techniques and geographic information system. Indian Institute of Remote Sensing report, pp. 1-76.

Roy, P.S., P.G. Diwakar, I.J. Singh and S.K. Bhan (1994d). Evaluation of microwave remote sensing data for forest stratification and canopy characterization. *Journal of the Indian Society of Remote Sensing*, 22: 31-44

Shirish, R.A., and P.S. Roy (1993). Landscape ecological analysis of disturbance gradient using geographic information system in Madhav National Park, Madhya Pradesh. *Current Science,* 68(3): 309-315

Shirish, R.A. (1994). Ecological analysis of vegetation from satellite remote sensing at Madhav National Park, Shivpuri, Madhya Pradesh. Unpublished doctoral dissertation.

Singh and P.S. Roy (1990). Growing stock estimation through stratified sampling on satellite remote sensing data. *Journal of the Indian Society of Remote Sensing*, 18: 29-42.

UNESCO (1973). International Classification and Mapping of Vegetation. Series 6. Paris.

Werle, D. (1989). Potential application of imaging radar for monitoring the depletion of tropical forests. In *Proceedings* of IGARSS 1989, Vancouver, B.C., Canada, pp. 1,383-1,413.

Wu, S.T. (1981). Analysis of results obtained from integration of Landsat multispectral scanner and Seasat synthetic aperture radar data. NASA-NSTL report No. 189, pp. 1-53.

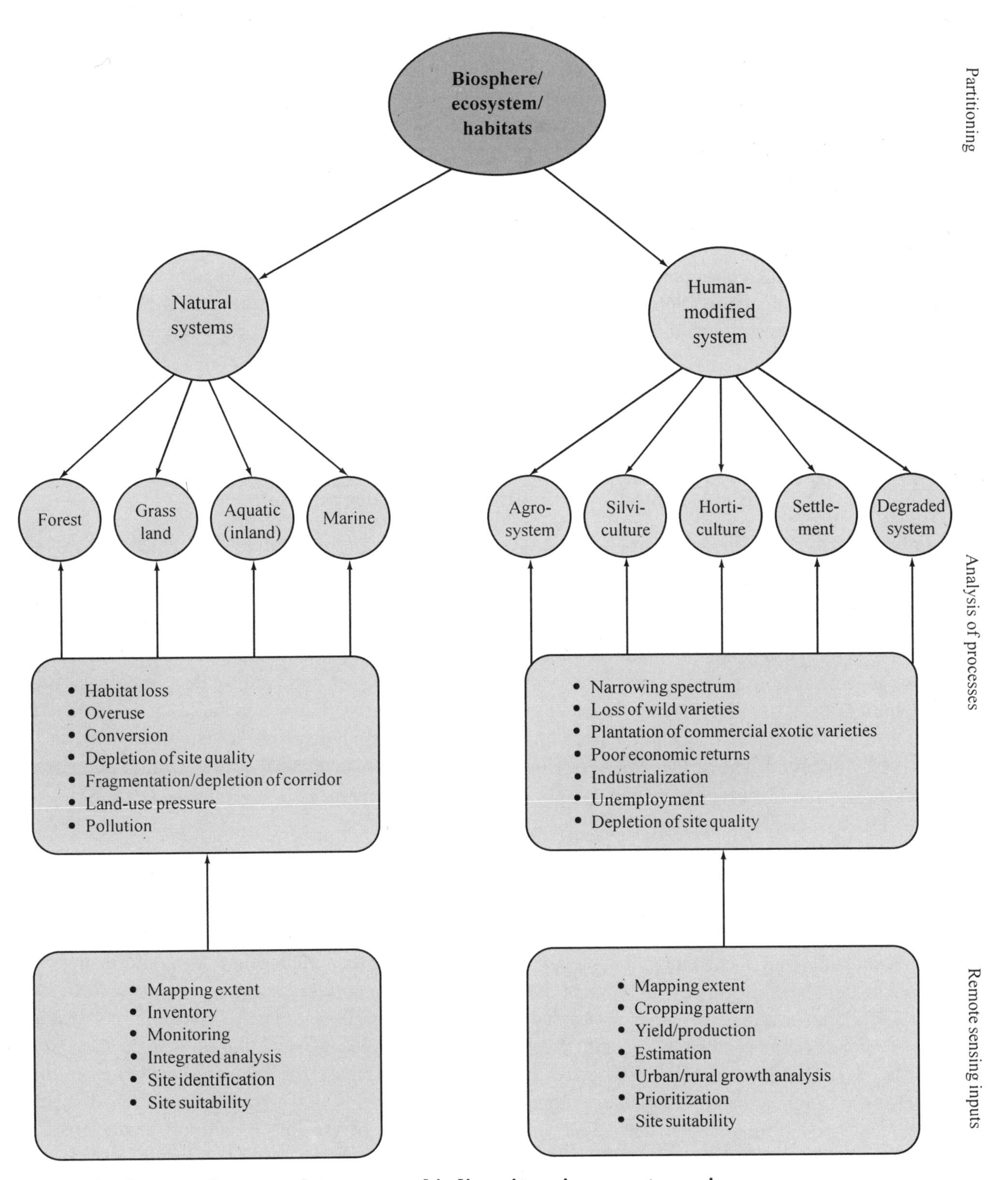

Figure 1. Suggested approach to manage biodiversity using remote sensing

REMOTE SENSING IN STUDIES ON LAND DEGRADATION

*R.S. Dwivedi**

A. Introduction

The importance of soil resources to human life needs hardly any emphasis since it is needed for agriculture, forestry, settlement, pastures, engineering, hydrology and many other fields. Continuous exploitation of soil resources to meet the demand for food, fuel and fodder for the rapidly growing human and animal population without providing adequate protection of the soil has led to the degradation of lands by way of soil erosion by wind and water, salinization and or alkalization, waterlogging, shifting cultivation and chemical pollution from effluents of the chemical industries.

In order to conserve our precious soil resources and to reclaim degraded lands, information on their nature, extent and physico-chemical characteristics is essential. Hitherto, this information has been generated mainly by conventional soil surveys using topographical sheets or cadastral maps as the database. With the development of aerial photo interpretation techniques in the late 1960s, the mapping effort was augmented appreciably. The launch of the Earth Resources Technology Satellite (ERTS-1), later renamed Landsat-1, in 1972 marked the beginning of the exploitation of satellite remote sensing technology for the management of natural resources. The ability of the Landsat series of satellites to provide a synoptic view of a fairly large area on a repetitive basis, and at regular intervals, and the amenability of these data to both digital analysis and visual interpretation make them a unique tool for natural resource mapping and management.

B. Mapping and monitoring land degradation

An estimated 175 million hectares of land, constituting 53.3 per cent of India's geographical area of 328 million hectares are subject to some kind of degradation. The degradation of land is brought about by a process classed as "land degradation". It is defined by FAO, UNEP and UNESCO (1979) as a "process which lowers the current and or potential capability of soil to produce goods or services". The processes recognized are (a) water erosion, (b) wind erosion, (c) salinization and/or alkalization, (d) physical degradation (loss of structure, sealing and crusting of top soil, reduction of permeability, compaction and so on), (e) chemical degradation (leaching of bases and build-up of elemental toxicity) and (f) biological degradation (loss of organic matter and decrease in microbial activities). All of these processes lower the soil's productivity, sometimes temporarily, often in a permanent or occasionally in an irreversible manner.

1. Soil erosion

Active soil erosion by water and wind accounts for erosion on over 140 million hectares of land in India. These lands are in different stages of degradation, namely those affected by sheet erosion, those by rill erosion, and those lands where gullies and ravines are formed. In order to reclaim them and bring them under the plough or put them to other appropriate use, information on the nature, extent and magnitude of erosion is of paramount importance. Remote sensing data, by virtue of synoptic coverage of large areas, have been found to be extremely useful in delineating eroded or ravinous lands, along with sand sheets and various types of sand dunes. Since the information on these lands is generally required on a watershed basis, the synoptic coverage of the terrain meets this requirement, wherein the boundaries of watersheds down to micro-watersheds could be delineated, and information on the extent and the magnitude of erosion, broad soil characteristics, land use and land cover could be generated. This information, along with the length and degree

* Land Degradation Division, National Remote Sensing Agency, Department of Space, Balanagar, Hyderabad, India.

of slope, and soil conservation measures employed, is extremely useful for watershed prioritization and subsequent treatment.

At the National Remote Sensing Agency (NRSA), Landsat MSS data has been used for qualitative assessment of soil erosion in parts of the Western Ghats, and four categories of erosion, namely "nil to slight", "moderate", "severe" and "very severe" were delineated. In another study, ravinous lands along the rivers Chambal and Yamuna in part of Uttar Pradesh, northern India, were mapped using Landsat MSS data, and three reclamative categories, namely "deep ravines", "medium ravines" and "shallow ravines" were delineated. The maps, thus prepared, have been found to be extremely useful in the land reclamation programme.

Moreover, under the Indian Remote Sensing (IRS-1A) satellite data utilization programme (IRSUP), soil erosion mapping was carried out in parts of Arunachal Pradesh, Manipur and Tripura states using Landsat MSS and TM data and IRS-1A LISS-I and LISS-II data. Both digital analysis and visual interpretation approaches were tried out. Based on broad soil characteristics as revealed in remote sensing data -- length and degree of slope, density of vegetation and the practice of shifting cultivation -- and soil conservation practices that were followed, (if any), four categories of erosion intensity units could be delineated: "nil to slight", "moderate", "severe" and "very severe". Besides, two categories of shifting cultivation areas (*jhum* lands) could also be delineated: "current *jhum* lands" and "abandoned *jhum* lands". The temporal nature of remote sensing data also helped studying the *jhum* cycle that is required for afforestation and rehabilitation of shifting cultivation areas.

2. Salt-affected soils

An estimated 7 million hectares of land in India are affected by salinity and/or alkalinity. These lands are invariably barren. For reclamation of such lands for agriculture, forestry or any other purposes, information on their nature, extent and magnitude is essential. As in the case of eroded lands, here also the space-borne multispectral data have been found quite useful in delineating such lands. Using Landsat MSS data, maps of salt-affected soils in parts of Punjab, Haryana and Uttar Pradesh have been prepared at 1:250,000 scale by NRSA. Besides, in an attempt to fully exploit the high spatial resolution of Landsat TM data, the magnitude of soil alkalinity (sodicity) was mapped in part of Mainpuri District of Uttar Pradesh, and two categories of sodic soils, namely "severely sodic soils" and "moderately sodic soils", could be delineated at 1:50,000 scale. The third category, i.e. "slightly sodic soils", could not be picked up on Landsat TM data as it was found to support crops and vegetation of varying density and vigour. Consequently, its spectral response pattern closely resembles normal soils, thereby making their delineation difficult. Besides, by making use of the temporal nature of space-borne multispectral data, the extent of salt-affected soils over a given period could be monitored.

Under the Remote Sensing Application Mission (RSAM), a project titled Mapping Saline/ Alkali Soils of India envisages the delineation and mapping of salt-affected soils of the entire country at 1:250,000 scale using Landsat TM data. Another project -- mapping the saline/alkali soils of Mainpuri District (Uttar Pradesh) and Ahmedabad District (Gujarat) at 1:50,000 scale through digital analysis and visual interpretation of Landsat TM data -- is nearing completion. After completion of this project we will be in a position to bring out the state-wise figures and total area under salt-affected soils in the country.

3. Waterlogged lands

Precise information on the extent of waterlogged land is lacking. Nevertheless, efforts have been made to work out the area of such lands based on sketchy information. An estimated area of 6 million hectares is subject to waterlogging in the country (National Commission on Agriculture, 1976). Of this, 3.4 million hectares are threatened by surface flooding, while on 2.6 million hectares, waterlogging is due to rising groundwater tables. Waterlogged lands that are wet at or near the surface are manifested as bluish green or greenish blue on standard false colour composite (FCC) prints. However, those lands where waterlogging is due to rising water table, the wetness is not manifested at the surface. As a result, it may be difficult to delineate them using remote sensing. Nonetheless, the vigour and density of vegetation may serve as an indicator for mapping such lands. Remote sensing data have been extensively used for mapping and monitoring waterlogged lands. In a recently concluded study, the temporal behaviour of waterlogged areas has been studied using time series space-borne multispectral data.

C. Conclusions

In order to meet the growing demand for food, fuel and fodder, available land and water resources need to be fully exploited, in accordance with their potential and limitations. Besides, degraded lands need to be reclaimed and put to proper use. It calls for timely and precise information on the nature, extent and physico-chemical characteristics of soils and the extent and magnitude of degradation. Space-borne multispectral data, by virtue of synoptic coverage of large areas, holds great promise in providing the desired information for soil resource management and for reclamation of degraded lands. Furthermore, the monitoring of degraded lands -- which aims at bringing out new areas that have undergone degradation or identifying degraded lands that have been reclaimed -- could be carried out using remote sensing data. While researchers are monitoring the degradation processes, variations in the spectral response pattern owing to change in the illumination geometry (sun azimuth/elevation angle), slope/aspect, moisture content and land cover need to be accounted for, and then the digital values may be converted into physical values like radiance and reflectance. Subsequently, these values can be related to changes that have taken place during the period in question.

The 23.5 m spatial resolution multispectral data from LISS-III and 5.8 m PAN stereo data from IRS-1C may help us derive information on the nature, extent, spatial distribution and temporal behaviour of degraded lands on a larger scale, which will enable better reclamation and management of these lands. What is more, the three-dimensional view of terrain provided by the PAN sensor may enable better appreciation of slope and aspect, which is required for soil erosion modelling.

References

National Commission on Agriculture (1976). Report of the National Commission on Agriculture, Part V, and Abridged Report, Ministry of Agriculture and Irrigation, New Delhi.

FAO, UNEP and UNESCO (1979). A provisional methodology for soil degradation assessment. FAO, Rome.

POTENTIAL EROSION MAPPING USING GIS TECHNIQUES IN DARAB WATERSHED, FARS PROVINCE, ISLAMIC REPUBLIC OF IRAN

*M. Morabbi, M.R. Varasteh, M. Ebrahimi and R. Haez**

A. Introduction

Soil erosion is a significant factor affecting the natural resources of land surfaces. In any watershed management plan the quality and quantity of erosion have to be considered. The lack of an adequate communication system makes ground surveys and conventional methods expensive and time consuming in mountain watershed studies, such as that in Darab, Fars Province, Islamic Republic of Iran.

Exposure of erodible marly, silty and saline formations such as Aghajary, Mishan, Pabdeh-Gurpi and salt domes, and the scarcity of dense vegetation cover in the Zagross mountains are major causes for a million tons of sediment yield annually. This has resulted in accelerated soil erosion, which in turn contributes to severe land degradation, heavy sediment load and flooding of areas downstream.

Several models can be applied for erosion level assessment in a watershed basin where there are not enough hydrologic data available. With the available data, the erosion potential model (EPM) is a well-known model that can be used for this study.

The objective is to generate a potential soil/rock susceptibility map of the Darab watershed basin. The Darab watershed was selected as the study area because it is currently experiencing dynamic developments. This paper may provide some useful information for planners of different projects intended for the area.

Darab watershed is a small basin located in the southern part of Fars Province (figure 1). The basin lies between latitudes of 28°30'N to 29°10'N and longitudes of 54°E to 55°E, with a total area of 314,852.7 hectares.

B. EPM method

The erosion potential method is an erosion model designed to predict erosion severity based on a specific land-use/land-cover management system. In this method, the erosion severity coefficient (Z) is calculated from the following formula:

$$Z = Y \times Xa (\phi + I^{0.5}) \quad (1)$$

where

"Y" is soil and rock resistivity to erosion, ranging from 2.0 for salt dome to 0.9 for limestone and dolomite (table 1);

"Xa" is land use coefficient, ranging from 1.0 for salt dome exposures to 0.05 for urban area (table 2);

"ϕ" is the soil erodibility factor, which defines the inherent susceptibility of different rocks to erosion. The value of ϕ for different rocks varies because of differences in significant rock properties such as mineral composition, texture, structure and permeability. "ϕ" represents the erodibility factor for the observed erosion processes in field observation or under a satel-

* IRSC, Remote Sensing Applications and GIS Department.

lite imagery study and ranges from 0.1 to 1.0 depending on the severity of present observed erosion (table 3);

"I" is the parameter for average land slope in percentage. Soil erosion is expected to increase with increases in slope gradient and slope length as a result of respective increases in velocity and volume of surface run-off. Furthermore, on a flat surface, raindrops splash soil particles randomly in all directions; on sloping ground, more soil is splashed downslope than upslope, the proportion increasing as the slope steepens;

"Z" is erosion coefficient calculated from the equation.

C. Methodology

Based on the EPM, the required information includes the following: land-use map, slope map and geological map. The land-use map of the study area was extracted from optical interpretation of Landsat TM imagery (FCC 432), path 161-40, acquired in 1990 (figure 2). The area is mainly covered with poor rangeland, agriculture and dry farming and some sparse forests. Different land-use classes are evaluated based on the influence of the erosion process, ranging from 0.05 to 1.0 (table 2).

From the geological point of view, Darab watershed is mainly composed of limestone, dolomite, alternations of calcareous sandstone, marl, siltstone and different Quaternary deposits (figure 3).

Based upon the resistivity of different lithologies, various deposits were classified into eight major classes, and an erodibility map was prepared from the original geologic map within the study area. Table 1 shows the erosion resistivity coefficient for each geological unit in the EPM (figure 4).

Table 1. Rock/soil resistivity coefficient to erosion in EPM (Y)

Bedrock	Mean value (Y)
1. Quaternary agriculture	1.7
2. Quaternary lowland	1.8
3. High terraces	1.4
4. Bakhtiary conglomerate	0.9
5. Aghajary formation (alternation of calcareous sandstone, marl, siltstone)	1.4
6. Tarbour-jahrom formation (limestone, dolomite)	0.9
7. Radiolarite	1.1
8. Salt dome	2.0

D. Spatial database construction

Digitization of all spatial features of the maps was done using PC Arc Info software and all different coverages were constructed. All editing was accomplished by using the ARC EDIT module. Before any analysis, the attribute data for coverages is built to represent the polygons. Data manipulation is accomplished by adding new items in tables for every coverage.

Slope coverage is subdivided into eight different classes ranging from less than 1 per cent to more than 60 per cent.

Land use coverage is subdivided into nine different classes, as shown in table 2 (figure 5).

Erosion resistivity coverage is grouped into eight classes ranging from 0.9 to 2, as shown in table 1.

Erodibility map coverage is graded into eight different values ranging from 0.1 to 1.0, based on table 3.

Table 2. Land use coefficient (X_a) in EPM model

Land-use classes	Mean value (X_a)
1. Range land	0.6
2. Agriculture	0.3
3. Swamp	0.8
4. Forest	0.6
5. Salt dome	1.0
6. Range land, dry farms	0.4
7. Saline land	0.4
8. Orchard, agriculture	0.4
9. Urban area	0.05

Table 3. Observed erosion coefficient values

Effective conditions on erodibility coefficient	Mean value (ϕ)
1. The whole area under severe gully erosion	1.0
2. Approximately 80 per cent of area under rills and gully erosion	0.9
3. Approximately 50 per cent of area under rills and gully erosion	0.8
4. Entire area under surface erosion and partially attacked by gullies and Karstic erosion	0.7
5. Entire area under surface erosion with no deep rills and gullying	0.6
6. 50 per cent of area under surface erosion, the remaining with no erosion	0.5
7. 20 per cent of area under surface erosion, remaining with no erosion	0.3
8. No erosion on land surfaces; only landslides occurred around river beds	0.2
9. No visible erosion on land surfaces, mostly crop fields cover the area	0.1
10. No detectable erosion on land surfaces; mostly covered by perennial crops and forest	0.1

Using the PC Arc Info overlay modelling technique, the above four layers (slope, land use, erosion and erodibility) were superimposed by UNION modules.

The result of the combinations is a new coverage with multiple polygons and an attribute table, as a product of the coverage data table.

A new item (Z) is added to the final table and its values are calculated according the following formula:

$$Z = Y \times Xa (\phi + I^{0.5}) \quad (2)$$

Newly generated polygons represent potentially susceptible areas for erosion. In order to classify the polygons in a meaningful manner for erosion severity mapping purpose, a new item is added as "erosion code". Polygons with $Z>1.25$ are classified as "severe erosion" and represented with a certain code number. The polygons with mean value of $0.85<Z<1.25$ are classified as "heavy erosion". The areas with $0.55<Z<0.85$ fall under "moderate erosion". Those polygons with $0.3<Z<0.55$ represent "low erosion", and, finally, polygons with $0.1<Z<0.3$ represent "very low erosion" areas (table 4).

Table 4. Erosion severity classification based on erosion severity coefficient (Z) in EPM model

Class number	1	2	3	4	5
Z ranges	$Z>1.0$	$0.7<Z<1.0$	$0.41<Z<0.7$	$0.2<Z<0.4$	$Z<0.19$
Mean value (Z)	1.25	0.85	0.55	0.3	0.1
Qualitative classification	Severe	Heavy	Moderate	Low	Very low

The final coverage shows erosion severity zones and can be mapped as potential erosion severity zoning (figure 6).

E. Conclusion

A geographic information system is a very useful tool in analysing large volumes of spatial data, to produce information that meets the user's need. Watershed management planning is a decision-making process for pursuing the maximum development with the minimum cost of financed resources. A GIS can provide a multi-criteria evaluation through overlay of maps containing relevant factors.

GIS may play a great role in the integration of data into a model framework. Furthermore, GIS can also provide necessary information to a model and has the capability of storing and displaying model output.

Erosion severity assessment using the GIS/EPM model results in high accuracy and speed. The acquired results from the EPM model can also be applied in erosion rate estimation with reasonable accuracy for rural development planning.

Bibliography

Bonham-Carter, Graeme F. (1994). *Geographic Information Systems for Geoscientists: Modelling with GIS.* Pergamon.

Brouwer, H.D. *Introduction to Geographic Information Systems.* ITC, Enschede, The Netherlands.

Land, L.J., et al. (1992). Development and application of modern soil erosion prediction technology: the USD experience. *Australian Journal of Soil Research,* No. 30.

Mahadavi, M. (1995). *Applied Hydrology.* Tehran University Press.

Saha, S.K., and L.M. Pande (1993). An integrated soil erosion inventory for environmental conservation using satellite and agrometeorological data. *Asian-Pacific Remote Sensing Journal,* 5(2).

(n.a.) (1986). *Soil Erosion and Conservation.* Longman Group Ltd.

Wischmeier, W.H., and D.D. Smith (1962). Soil loss estimation as a tool in soil and water management planning. International Association of Science Hydrology publication, No. 59.

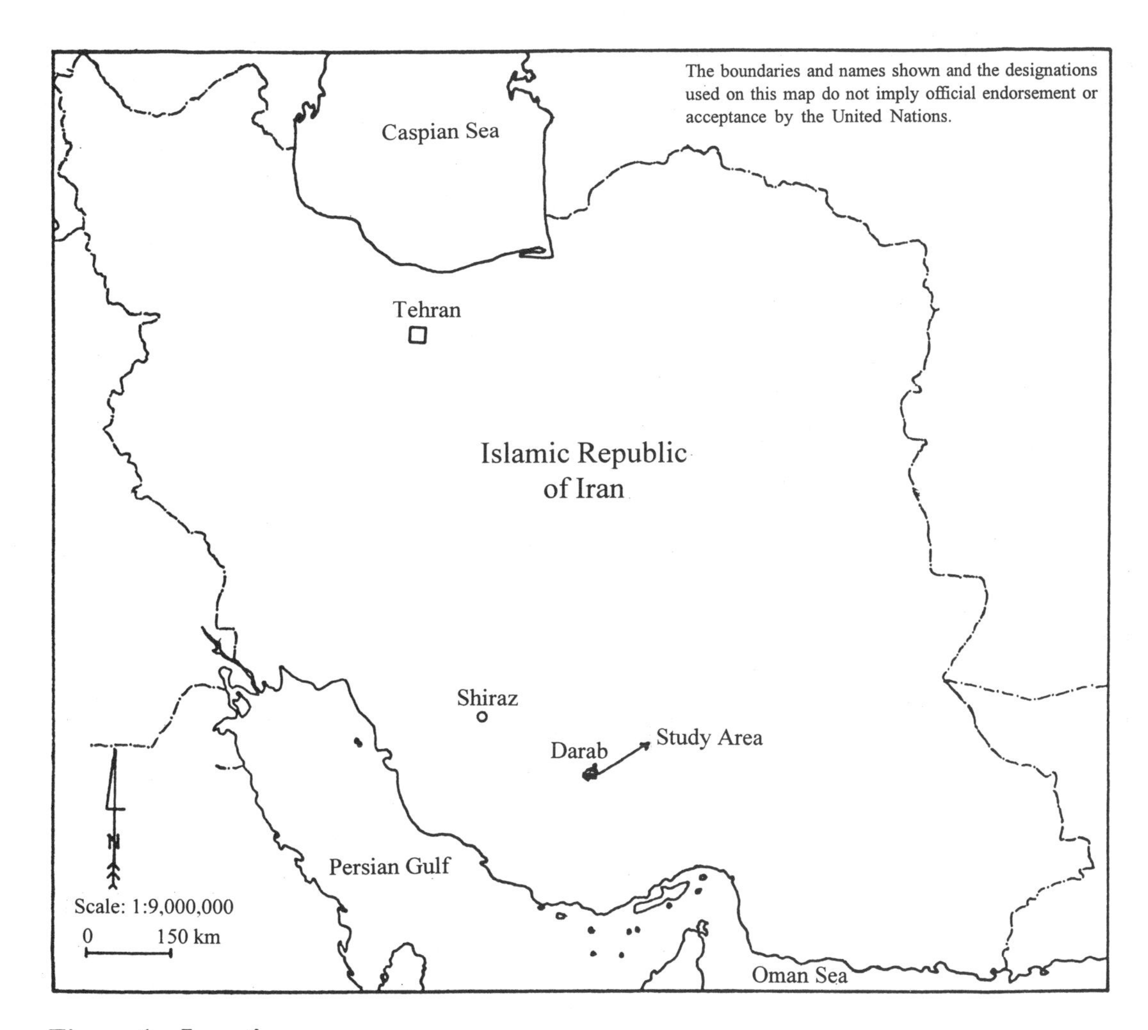

Figure 1. Location map

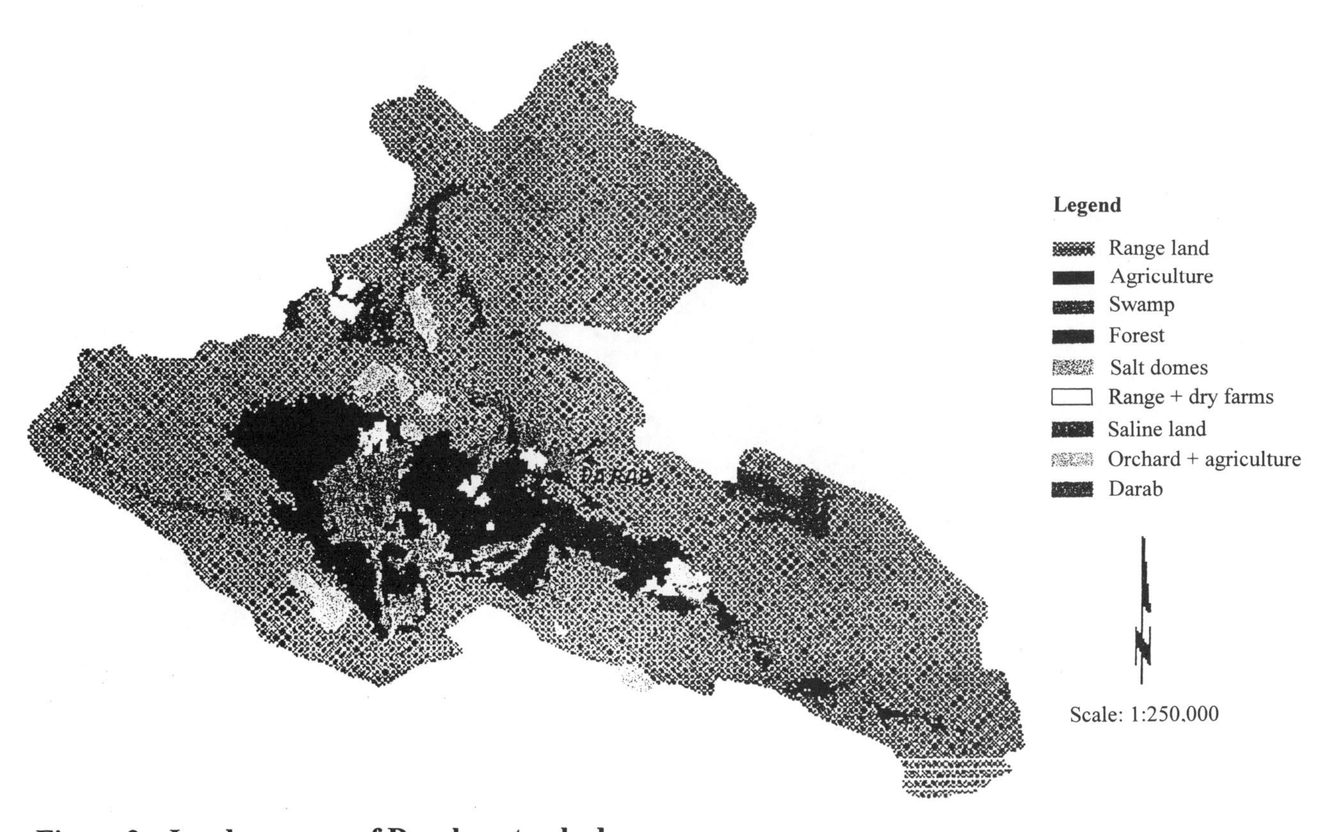

Figure 2. Land-use map of Darab watershed

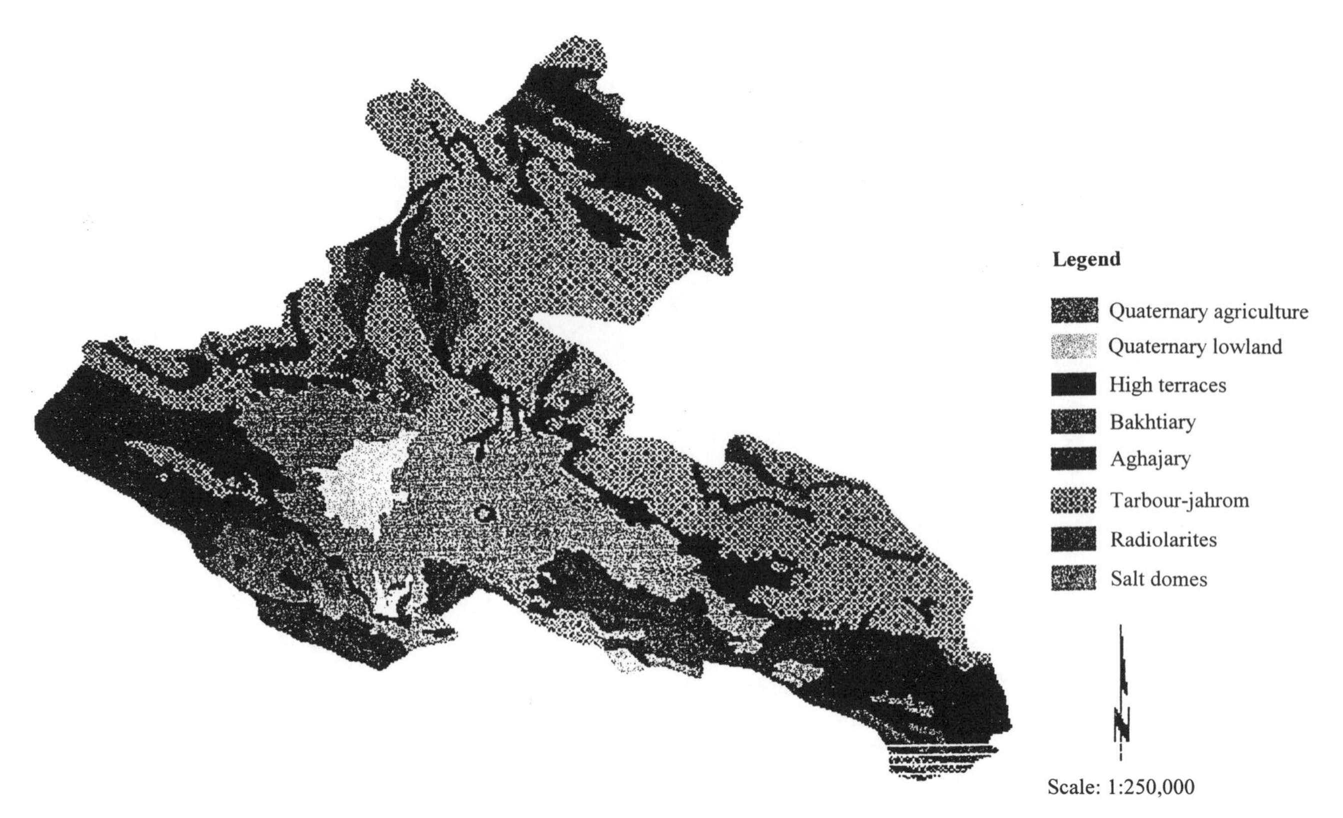

Figure 3. Geological map

Figure 4. Erodibility map

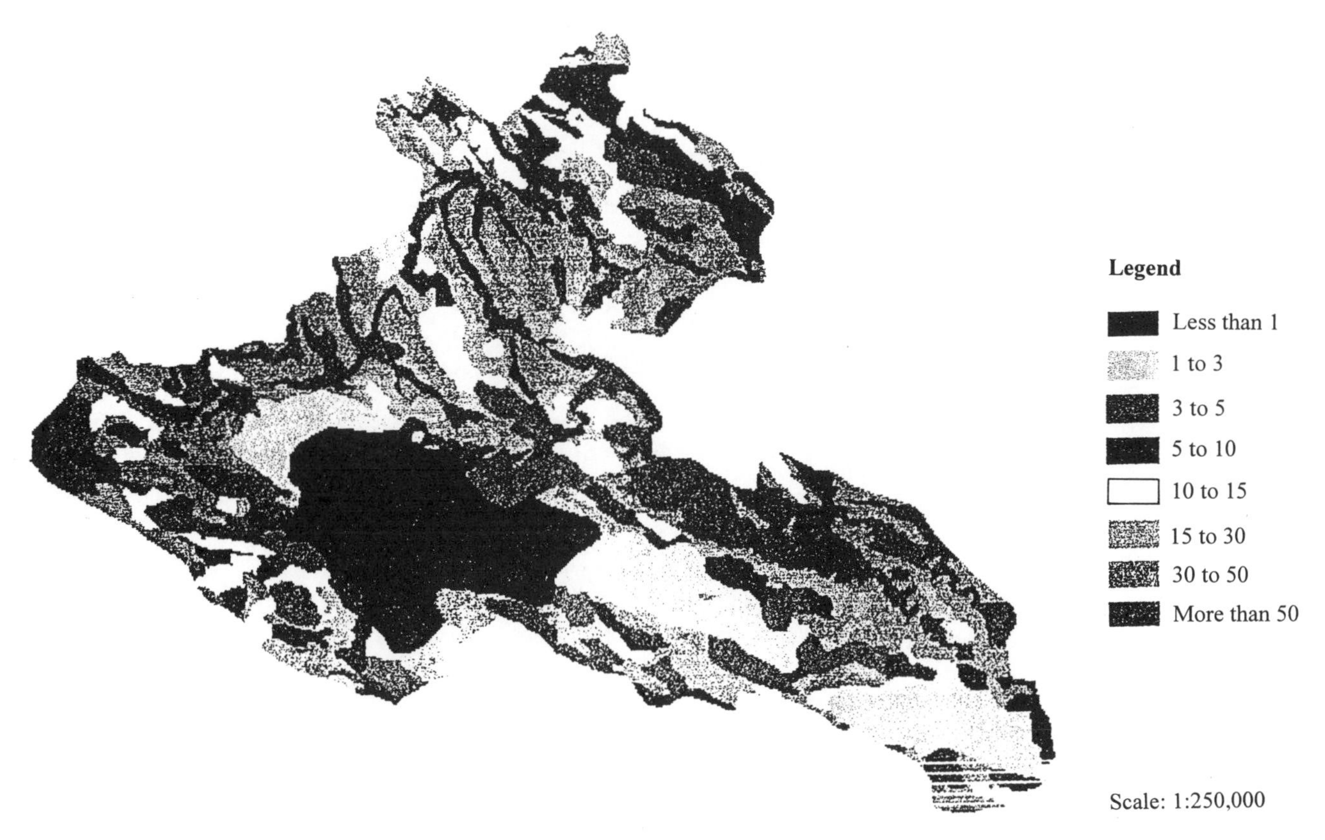

Figure 5. Slope map

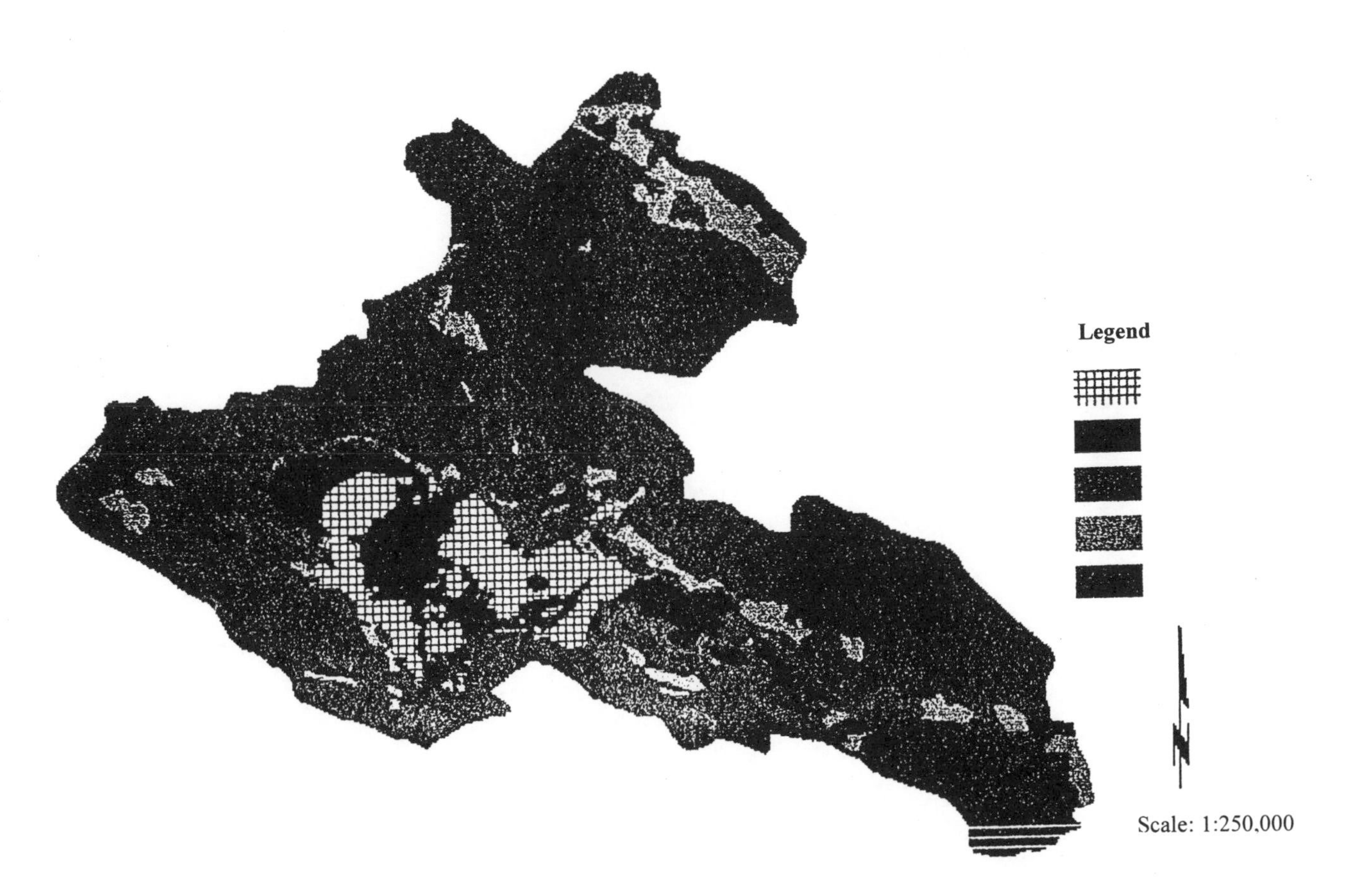

Figure 6. The erosion map

PART SEVEN

WATER RESOURCE MANAGEMENT AND FISHERIES

WATER RESOURCE MANAGEMENT AND FISHERIES APPLICATIONS

*A.T. Jeyaseelan**

A. Introduction

Resulting from the population explosion and accelerating threats to agriculture in some regions caused by deforestation, soil erosion, declining soil fertility and water resources, the application of the new advanced techniques such as remote sensing and GIS to monitor and manage the natural resources is essential in the developing countries.

The key to sustainable development and improved quality of life in many parts of the world is the availability of fresh water resources. Water is critically scarce in 22 countries where the renewable water resource is less than 1,000 cubic metres per person. In South Asia, of which India is a part, the per capita use of water is about 4,200 cubic metres. However, the expected increase in global population by more than 50 per cent over the next 30 years and increasing urbanization will place immeasurable stress on available water resources in rural areas.

While the total availability of water over South Asia presents a comforting picture, its spatial and temporal distribution, however, is incompatible with the need. Furthermore, increasing the water supply to towns and industries, coping with natural disasters such as drought and flood, maintaining acceptable water quality for human and other uses, designing and operating conveyance systems, all these and more must be based on reliable and objective assessment of quantity and quality of water resources. Space technology, in consonance with geographic information systems (GIS), can play a critical role in meeting these information needs for planning, monitoring and improved management operations of water resource systems. In addition, remote sensing plays a major role in identifying rural problems by the planners and managers for sustainable management of water resources.

B. Water resource management

Water resource management essentially addresses a linkage between availability and demand, in which both quality and quantity are considered and both conservation and control are addressed. The water demand that needs to be satisfied can be consumptive use or non-consumptive use. Conjunctive use of surface and groundwater is an integral part of water resource management. The development and management of water resources also involves soil, topography and land use management.

The conventional hydrologic measurements on the ground suffer from limitations of reliability, time effectiveness and adequacy. These measurements are also discrete in space, necessitating areal averaging methods. Repetition of ground measurements many times is rarely done because of the constraints of manpower and funds. Measurements over inaccessible areas and inhospitable terrains are also limited. It is in this context that remote sensing techniques along with GIS can complement and supplement ground measurements to enable collection of data needed for sound water resource management.

C. Remote sensing

Currently, aircraft and spacecraft employ a wide range of sensors for collection of water resources information. Metric camera systems, multispectral scanners and passive and active microwave systems can be installed on board aircraft, to fly at different altitudes, to provide detailed water resources information. Polar orbiting satellites such as Landsat, SPOT and IRS have scanners

* Water Conservation, Water Resources Group, National Remote Sensing Agency, Hyderabad, India.

to provide panchromatic and multispectral information of varying spectral, spatial, temporal and radiometric resolution. The NOAA series of satellites, in addition to weather-related data, provide terrain-related information for use in snow studies, sea surface temperature and vegetation monitoring. The geostationary INSAT satellite provides operational information on weather every three hours.

Satellites like ERS-1 and Radarsat carry microwave instruments such as the altimeter, scatterometer and synthetic aperture radar to study ocean- and land-related phenomena. Although IRS-1B, launched in 1991, is identical to IRS-1A, the third satellite in the series, IRS-1C, and its follow-ons carry sensors of improved spatial resolution, wide field sensing capability and stereo coverage range. SPOT-4 will carry a "vegetation instrument" for global vegetation monitoring. Landsat-7's Enhanced Thematic Mapper will have improved spatial resolution, panchromatic band and selective mode of transmission.

D. Applications of remote sensing in water resource management

Sustainable water management involves accurate assessment of water availability in surface and subsurface storage. This includes assessment of snow cover/precipitation and the volume of snow-melt run-off likely to occur in the river and reservoir storage systems. The effective monitoring of available surface storage through remote sensing is of vital importance to hydroelectric power generation, irrigation and drinking water supply. The other activity in water management is watershed prioritization for improved quality and quantity of water. Remote sensing and GIS play a major role in watershed prioritization by assessing current soil, slope, precipitation and land cover information. Water management activities for sustainable development also include proper distribution of water based on the need for irrigation, drinking water, agriculture and industries. Remote sensing and GIS, with demographic details, provide a way to estimate the correct requirements and ensure proper distribution by regular monitoring. Finally, to cope with natural disaster management related to water such as floods and drought, which are frequent in this part of world, remote sensing is used for pre-disaster planning, early warning by monitoring and for factual impact assessment. In this paper, various case studies are presented, as Indian examples of sustainable water management.

E. Rainfall monitoring

Several aspects of rainfall hydrometeorology are now amenable using data from satellites. They include (a) mapping the areas likely to receive rain, (b) mapping total possible rainfall, (c) assessing extreme rainfall events and (d) forecasting rainfall. The best method for rainfall mapping would be based on an active microwave (radar) system. In the absence of radar in most of the satellites, indirect methods are used relating ground-observed rainfall and satellite-observed radiances from visible, infra-red and microwave regions of the electromagnetic spectrum. The rain-bearing clouds can be identified from satellite data due to their brighter or higher radiance in the visible region and low cloud top temperatures in the infra-red region. Many methods or models are available to provide good rainfall estimates from the geostationary type of satellites such as INSAT and frequent revisit polar orbiting satellites such as NOAA.

F. Snow-melt run-off

Snow acts as a reservoir, so assessing its possible effect on snow-melt is essential for planning and management of various multi-purpose projects on snow-fed rivers. Unlike the conditions in many other countries, where snow fields stretch along flat terrain and mountains that are easily approachable, the snowbound land in India is mostly in difficult and inaccessible mountainous terrain. Therefore, remote sensing remains the only practical way of obtaining snow information.

Snow, by virtue of its high reflectance, can be easily detected or identified on any visible or near-infra-red image, and fresh snow has higher reflectance than old snow has. The temporal attribute of satellite imagery would be useful in tracing the changes in the albedo of the snow pack as well as total depletion of the snow cover due to rising temperatures. The principal characteristics of the process of the evolution of the snow cover is its density change and the settling of individual layers, which are to a great extent influenced by the temperature region of electromagnetic radiance. The temperature of snow during the day and melting period stays at 0° centigrade. Subtle changes in the reflectivity of snow can be used to estimate snow depth up to 15 cm. For depths greater than this, sensors other than optical sensors should be used.

The Himalayan rivers are snow-fed, and they offer large hydropower potential on account of their topography and the perennial nature of the flow. Snow-melt run-off estimation is therefore a critical item for planning and management. National Remote Sensing Agency (NRSA) studies in the Sutlej basin have shown very successful results in accurately estimating seasonal snow-melt run-off into the Bhakra reservoir. This model has been improved to provide revised/residual forecasts through the months of April, May and June. The snow-cover run-off model (SRM) in the Parbati River at the Pulga dam site and Beas River at Thalot has provided seven-day forecasts through the March-July period. Improvement on short-term snow-melt run-off, necessary for reservoir operation, are in progress and are being extended to other Himalayan basins.

G. Rainfall run-off

Rainfall run-off modelling in Gurpur and Sitanadi basins of Karnataka State has integrated the SCS model with remotely sensed information on hydrologic land-use. Use of ERS-1 SAR data for delineating high soil moisture zones (which significantly contribute to run-off) is being explored. Urban run-off modelling has been attempted to provide valuable inputs for effective drainage planning.

H. Surface water storage

Surface water storage is the primary input for agriculture by way of major, medium and minor irrigation. For a predominantly agricultural country like India, the uneven distribution of rainfall, even during the monsoon, means that effective surface water storage and monitoring of it are necessary. It involves inventory location identification, contributing watershed and assessing water storage.

Delineation of water bodies and determination of water-spread area at different times of water accumulation and water depletion periods can be carried out through sensors operating in the near-infra-red and microwave regions. The reason is that water totally absorbs near-infra-red radiance, so the land-water contrast is strong. Remote sensing measurements for estimating water depth and variation in bottom topography can be obtained in blue and green wavelengths, which facilitate greater penetration in clear and turbid water. Accurate profiling of water depth can also be accomplished by LIDAR (laser radar) systems. High-resolution satellite data, along with ground-observed depth measurements, have been used to assess storage in most of the major reservoirs in the country.

I. Irrigation management

Nearly 85 per cent of the current utilization of water in India is for agriculture, even though hardly 9 per cent of the annual precipitation of 400 million ha-m is utilized for irrigation. By the turn of the century, 225 million tons of food grains will need to be produced to feed more than a billion in population, calling for a 6 per cent annual growth rate, double that achieved during the "green revolution". The dramatic increase in food production in the post-independence era has to a large extent come through large-scale irrigation, increasing from 20 per cent in 1953 to 35 per cent of total arable land in 1990. At the global level, 40 per cent of the world's food comes from irrigated land, and it is estimated that half to two-thirds of the increase in food production in future will have to come from irrigated land. However, the economic costs of new irrigation can be expected to double or triple existing costs, making it difficult to sustain irrigation. The technology should thus shift from irrigation development to more efficient irrigation management. The performance of irrigated systems, however, has generally been below expectations, with low economic and financial returns discouraging further investment. It is estimated that even if a marginal 2 per cent improvement can be achieved in the operating efficiency of existing projects, it would mean an additional irrigation potential of 0.5 million ha. Adverse impacts of salinity and waterlogging as a result of over-irrigation and inadequate drainage are also limiting factors. Growing environmental concerns with large projects is also another constraint. Other users are also likely to compete to the disadvantage of allocation of water for agriculture. Therefore, increased food production will have to come with less water but through increased cropping intensity and higher yields.

Multi-year satellite data has been used to evaluate the performance in many irrigation systems across the country. The anticipated increase in irrigated area, equitable distribution and crop productivity under programmes such as the centrally sponsored CAD scheme and the National Water Management Project have been studied in the Bhadra project, Malampuzha project, RDS scheme, Salandi, Hirakud and Mahanadi Stage I and II projects. The temporal and spatial analysis of satellite data have indicated problem pockets of poor performance. Diagnostic analysis, supported by farmer surveys, has identified causative factors for corrective management. Satellite data has helped in these projects in identifying whether the recommended cropping pattern has been realized, thus providing input for changes in policies and operational plans. Spatial analysis of paddy transplantation periods of the Bhadra and Hirakud projects have raised policy issues of relevance to irrigation scheduling, canal maintenance and agricultural productivity. Satellite-derived paddy acreage and condition data have been used to improve the design of crop cutting experiments in paddy irrigation systems to provide more reliable crop yield estimates. Satellite data has also been analysed to map the current status of waterlogging and soil salinity in Sarada Sahayak, Nagarjunasagar and other irrigation projects. Analysis of satellite data has also helped to evaluate effectiveness of reclamation programmes by monitoring the extent and severity of soil limitations through the years. Conjunctive surface and groundwater utilization in irrigation water management is helped by water budgeting methodology, with satellite- and ground-derived inputs, as in the thirty-sixth distributary of Sriramsagar project in Andhra Pradesh State. Effective water balance studies can also be helped by indirect satellite identification of significant return flows as in RDS project in Andhra Pradesh State and Bhadra project in Karnataka State.

Creation of new potential is being helped by generation of basic land-use and soil maps from satellite data, as well as derived land capability and irrigability maps. Close contour information is generated through photogrammetric measurements of large scale aerial photographs and high-resolution satellite imagery from SPOT and IRS-1C satellites.

The Food and Agriculture Organization of the United Nations (FAO) estimates that an average of only 45 per cent of water released from reservoirs actually reaches the crop. Satellite data has helped identify canal reaches that require lining. A factor in irrigation is the large extent of unauthorized irrigation in many projects and poor recovery of water-use rates. Satellite data in RDS project and Nagarjunasagar project in Andhra Pradesh have delineated the extent of unauthorized and under-reported irrigated areas. A quick computation indicated a loss of Rs 7 lakhs (Rs 700,000) to the exchequer as a result of under-reporting in RDS scheme. The satellite study on the left bank canal of Nagarjunasagar project revealed significant under-reporting of irrigated acreage, particularly under tanks in Zone III, which are supplemented by canal water but not reported.

Satellite remote sensing techniques have been applied even to small tank irrigation schemes. A study with multi-year satellite data in Kattiampandal tank in Tamil Nadu State has studied how farmers respond to differing rainfall patterns and inflow into the tank. A reappraisal of tank storage through multi-date satellite data in Madurantakam tank in the same state helped in planning optimum irrigation utilization.

Functional and dysfunctional tanks have been inventoried using satellite data. Foreshore encroachment and unauthorized utilization of tank beds have been detected. Silting status of tanks can be studied from analysis of catchment characteristics, water quality and reduction in registered *ayacut*.

J. Watershed management

Inappropriate land-use practices in the upstream catchment can lead to accelerated erosion and consequent silting up of reservoirs. Watershed management is thus an integral part of any water resources project.

The prioritization of watershed is based on sediment yield potential so that the treatment would result in minimizing sediment load into the reservoir. Satellite data has been extensively used in many project through the sediment yield index (SYI) method developed by All India Soil and Land Use Survey. Sediment yield prediction models have been used to provide quantitative silt load estimates in Jurala project and Asan and Ukai watersheds. Multi-year satellite data is also used to monitor the impact of watershed management programmes.

Remote sensing provides information for use in many empirical models developed elsewhere and in India. While simulation models such as the Colorado State University model are more rational and physical-based, calibration and use of such models have limited remote sensing inputs.

Watershed management for soil and water conservation is an integral component of the Integrated Mission for Sustainable Development (IMSD) programme taken up in 157 districts of the country. Implementation of appropriate rainwater harvesting structures in selected watersheds under this programme has demonstrated significant benefits by way of increased groundwater recharge and in agricultural development of once barren areas.

K. Reservoir sedimentation

Many reservoirs built at huge investment are undergoing rapid silting and loss of storage capacity and consequent reduction in their economic life. It is estimated that 20 per cent of the live storage capacity of these reservoirs will be silted up by 2000 AD, resulting in an average loss of 60,000 ha of irrigation potential every year. Conventional hydrographic surveys to reassess reservoir capacity are time-consuming and costly. Satellite remote sensing techniques have been demonstrated to be cost- and time-effective. Multi-date satellite data have been used as an aid to capacity surveys of many reservoirs such as Hirakud, Nathsagar, Ujani, Tungabhadra, Malaprabha, Ghataprabha, Osmansagar, Nizamsagar, Himayatsagar and Sriramsagar. While this technique helps in revising the capacity table between minimum and maximum draw-down level observed in satellite data, the loss of dead storage capacity cannot be estimated except qualitatively. Realistic appraisal of reservoir capacity will lead to appropriate utilization plans.

Satellite data has also been related to surface concentration of suspended sediment in many reservoirs, providing information on the sediment distribution, circulation pattern and active silting zones.

Satellite comparison of turbidity levels in irrigation tanks can identify those receiving large silt inflows, for initiating desilting operations or catchment treatment.

L. Water quality

Increasing environmental concerns about deteriorating water quality are not well supported by ground monitoring mechanisms. Remote sensing of water quality can complement ground efforts in mapping and monitoring point and non-point pollution sources, the influx and dispersal of pollutants in the aquatic environment and consequent impact such as algal bloom and weed growth.

Point source identification calls for high-resolution satellite data. Regional models of non-point source pollution loading, as in the case of water supply reservoirs in Hyderabad, will benefit from remote sensing inputs on land cover and land use, supported by sample ground data collection. Aerial surveys with multispectral scanners have demonstrated how the pollution influx and dispersal in riverine (Godavari River) and lacustrine (Hussainsagar) environments in Andhra Pradesh could be monitored. Salinity intrusion into Hooghly estuary have been aerially surveyed and studied.

Growth of aquatic weeds and algal bloom, as indications of eutrophication, have been mapped from satellite and aircraft data. In general, remote sensing techniques can be successfully applied in all environments where there is a change in colour, temperature or turbidity. Care needs to be taken to support statistical models of remote sensing of water quality with an understanding of physics. Ground truth requirements are also more stringent than in land remote sensing. The GIS technology provides enhanced capability for water quality modelling.

Satellite data have been used in river action plans such as in the Yarnuna River to identify sites for locating sewage treatment plants.

M. Groundwater

Nearly half the current irrigation potential of 80 million ha and large urban and rural populations draw from groundwater resources. Studies have shown how integration of geological, geophysical and remote sensing data in geostatistical models can help improve the success rates of even high-yield irrigation wells. This would not only help in providing much-needed water to thirsty farmlands but would also help to obtain full reimbursement by funding agencies such as NABARD.

According to recent reports only 73 per cent of India's population have access to safe drinking water. Under the national drinking water technology mission, covering more than 600,000 villages, satellite data was interpreted into hydromorphological maps in 1:250,000 scale for the entire country. These maps provide input to problem villages to aid in finding primary sources for drinking water supplies. Based on statistics of over 170,000 bore wells sited with aid from remote sensing, a success rate of about 90 per cent has been observed in groundwater targeting, compared to a 45 per cent success rate using conventional procedures. More detailed maps at 1:50,000 scale have been produced for critical areas such as towns and problem villages. A statistical groundwater model is also under development to quantify satellite-derived groundwater potential. Remote sensing techniques have also helped in identifying recharge sites for groundwater augmentation so that sustainable yield can be maintained.

N. Flood management

More than one-eighth of the country is flood prone and one-fifth of this area experiences floods in any one year. Recent years have seen major floods in the Ganga and Brahmaputra river systems, in Punjab, in Orissa and in Andhra Pradesh states. Structural and nonstructural flood control measures have decreased flood damages in selected areas.

The large area synoptic coverage from satellites has helped to map flood-inundated areas and estimate flood damages for more than a decade now. Soon after the occurrence of a significant flood event, the inundated area is mapped and damage statistics are generated and sent to concerned state and central government agencies. Flood mapping has been operationalized since 1986. Satellite remote sensing and geographic information system techniques have been integrated in a case study in Assam State to provide spatial information at the mouza level on flooded areas and damage to croplands and facilities such as roads and railroads.

While the utility of optical remote sensing data can be limited by cloud cover, microwave remote sensing data has been used from 1993 to penetrate cloud cover and see the ground. India may be the first country to operationally use microwave data for flood mapping.

The Global Positioning System (GPS) is being used to aid development of digital elevation models (DEM) of a flood prone area in Andhra Pradesh State to enable assessment of spatial inundation at different water levels in the river. When the satellite-derived land cover/use and ancillary ground-based socio-economic data is draped over DEM, flood vulnerability can be assessed to provide location-specific flood warnings. Remote sensing information inputs are being evaluated for integration with existing flood forecasting models. A study in the lower Godavari basin in Andhra Pradesh State will integrate SRS-GIS-GPS technologies for improved flood forecasting and spatial flood warning. This cooperative effort between NRSA, the Central Water Commission and the government of Andhra Pradesh will be the forerunner for similar efforts in other flood-prone river systems.

Preliminary flood risk zone maps using multi-year satellite data have been generated for Kosi, Jhelum and Brahmaputra rivers. A joint project between the Central Water Commission and NRSA will focus on flood management of the Yamuna River upstream of Delhi.

Structural flood control measures such as embankments and spurs have been mapped along with current river configurations using high-resolution satellite data to identify vulnerable reaches. Such annual mapping efforts in Kosi Gandak, Ravi and Sutlej rivers have helped in undertaking engineering measures to strengthen vulnerable reaches or to plan new structures. The effectiveness of river training works could be studied by monitoring river morphology.

In the context of reawakened interest in the safety of large dams, high-resolution satellite data of downstream areas can be analysed to provide information on vulnerable areas requiring protection.

O. Drought monitoring and management

India's economy is largely dependent on agricultural production. Owing to abnormalities in the monsoon precipitation in terms of spatial and temporal variations, especially with the late onset of the monsoon, droughts due to prolonged breaks and early withdrawal are a frequent phenomena over many parts of the country. Thirty-three per cent of the area receives less than 750 ml of rainfall and is chronically drought prone. Thirty-five per cent of the region receives 750-1,125 mm; these areas are also subjected to drought once every four to five years. Therefore, 68 per cent of the total sown area, covering about 140 million ha, is vulnerable to drought conditions.

Agricultural drought monitoring should be concerned with crop condition assessment in terms of crop stress during its growth and the yield prediction; estimates should be based on the crop condition with respect to crop types. Stress may be defined as any factor that reduces the productivity of the canopy below its potential or optimum value. Spectral indices have been used for some time for monitoring of vegetation by remote sensing. The original indices were based on a combination of visible and near-infra-red bands, although other techniques have recently been proposed using microwave backscatter. Much development work at the moment centres around how vegetation indices should be constructed and interpreted. In order to assess the stress, the set of vegetation indices should enable researchers (a) to measure canopy density or light absorption for photosynthesis, (b) to establish the presence of significant stress, (c) to distinguish different classes of stress and (d) to measure the degree of stress and its effect on productivity. The conventional two-band vegetation index (VI) such as IR/R or the normalized difference vegetation index (NDVI) can only meet the first of these requirements, subject to the limitations explained earlier. The other requirements are derived from the experience of vegetation indices with known ground conditions of previous years.

Whereas the NOAA satellite covers India twice a day (and there are two such satellites), satellites such as the IRS (Indian Remote Sensing) satellite, Landsat or SPOT revisit the same area every 22 days, 16 days or 26 days respectively. However, NOAA satellite's AVHRR (Advanced Very High Resolution Radiometer) sensor has a spatial resolution of 1.1 km at nadir, and the IRS WiFS sensor has a resolution of 188 m compared to 36.25 m to 10 m from other satellites. Since countrywide monitoring capabilities require frequent revisits of a satellite more than spatial resolution, the NOAA satellite is the primary satellite for the national drought surveillance system. Once early warning of the drought-affected areas and severity has been provided by the national system, a closer, detailed survey of affected areas would come from Earth resource satellites.

The National Agricultural Drought Assessment and Monitoring System (NADAMS) has been providing since 1989 biweekly drought bulletins through the kharif season (June to December) for the eleven states of Tamil Nadu, Andhra Pradesh, Karnataka, Maharashtra, Orissa, Gujarat, Uttar

Pradesh, Haryana, Madhya Pradesh, Rajasthan and Bihar. Bulletins describing prevalence, relative severity level and persistence through the season are issued by district level officers, such as the collector and district agricultural officer. The drought assessment is based on a comparative evaluation of satellite-observed green vegetation cover (both area and greenness) of a district in any specific time period with the vegetation cover of similar periods in previous years. This comparative evaluation helps in fixing the current season in the scale of historical agricultural situations. NADAMS currently uses a district-wise database of VI extending from 1986. During the 1990 kharif season, drought assessments were sent to drought-affected states and districts within 48-72 hours of every biweekly period by telex or telegram, and printed bulletins were dispatched within 10 days. This nationwide service has been found to be useful for providing first cut alert of drought conditions. During the continuation of project, concurrent with the ninth five-year plan period of 1997-2001, quantitative estimates of drought impact on crop production will be the major objective. Spatial variability of drought within the district will be addressed. The ten years of data on satellite-derived vegetation indices, coupled with ancillary ground agricultural information, are currently being analysed to study rainfall-vegetation index relationship, spatial coherence of drought conditions and drought vulnerability.

P. Conclusions

While remote sensing applications to water resource management have made considerable progress, still more remains to be done to operationalize remote sensing applications. Sustained efforts towards consolidating results obtained so far and in developing operational methodology packages are needed. The need for the day may be formulation of a national project on river basin development in which all aspects of water resource management could be integrated using remote sensing techniques (and ground methods) to develop operational methodologies, and to integrate such procedures in the day-to-day working of water resource organizations. Remotely sensed data should not be exotic tools to water resource scientists and engineers. Remote sensing techniques should be routinely used in feasibility/reconnaissance investigations by water resource organizations. Monitoring of water projects in terms of adverse impacts is another area that requires operational use of remote sensing techniques. Progress in these operational applications will certainly lead to sustainable water management in the developing countries with exponentially growing populations.

REMOTE SENSING FOR OCEAN RESOURCES

*A.N. Nath**

A. Need for remote sensing of oceans

Oceans are governed by physical, chemical, biological and geological factors. As does the land, oceans also possess vast living and non-living resources. Owing to the scarcity of the resources on land, one should look for the wealth of ocean resources and find out methods of exploitation. Remote sensing plays an important role in locating and exploiting fishery resources. At present, marine fish catches along the Indian coast are estimated at about 2.2 million tons per annum, out of the estimated potential of 4 million tons. Temperature and ocean colour are useful in locating marine fishery resources. The sea surface temperature (SST) sensor in NOAA-AVHRR (the National Oceanic and Atmospheric Administration's Advanced Very High Resolution Radiometer) has been extensively used in locating the potential fishing zones (PFZs) in near-real time by many developed countries. A national project on application of remote sensing for the identification of PFZs along the east and west coasts of India using NOAA-AVHRR-derived SST information has been carried out by the National Remote Sensing Agency (NRSA). The procedures of the preparation and dissemination of the data, the collection of fish catch feedback data, and the validation of the satellite data with the sea truth data is discussed in detail in this paper. Fish catch feedback data received from fishing centres indicate three to four times higher catch in the identified PFZ areas than in the non-PFZ areas.

The state of the sea is governed by a variety of factors. Physical factors include sea surface temperature, surface winds, tides, waves, currents, eddies, internal waves and circulation patterns. There are chemical factors like salinity, oxygen, water quality, dissolved gases, fertility and contamination of sea. Biological factors are phytoplankton. zooplankton, fishes and various other living resources. Geological factors constitute the non-living factors, such as topography, seabed variation and sea mounts, to name a few.

Since oceans are vast and associated with immense economic benefits, they need to be monitored repetitively on a spatial and temporal basis. Remote sensing of oceans can be useful in many ways: (a) it can give a global picture of oceans for broad study of basin-wide phenomena; (b) it can observe regions that are inaccessible or are not easily studied by ships, like certain regions around Antarctica; (c) it can make various measurements of ocean waves, oceanic rainfall and winds, which cannot be recorded by ordinary means; (d) it can boost fish production by helping in prediction and forecasting of marine fishery resources; (e) it can help in Earth climate studies through regular monitoring of the environment; and (f) it can help with weather forecasts, which are routinely made with the help of satellite remote sensing data, so that instructions and warnings regarding storm or cyclones being can be disseminated to the fishermen and ocean-going ships in time.

B. Need for ocean resource surveys

The ocean is not a homogeneous medium, although it is a single-phase environment. There are different types of resources in the water column as well as on the seabed. Although India's oceans are rich with living and non-living resources, ocean resource studies still have much progress to make. Central government organizations such as the Fishery Survey of India, Central Marine Fisheries Research Institute and National Institute of Oceanography are working on the ocean resources of India.

* National Remote Sensing Agency, Balanagar, Hyderabad, India.

C. Scarcity of resources on land

Because of urban sprawl and deforestation, land resources are diminishing day by day. Uncontrolled population growth around the world and especially in India is an important factor in the depletion of the resources on the land. The restricted availability of the land for increasing land-bound production, coupled with the growing demand for protein-rich food, has caused worldwide concern to find ways to enhance the harvest of resources available in the sea.

D. Wealth of living resources in the ocean

At present the marine fish catch along the Indian coast is estimated to be about 2 million tons, as against the estimated potential of 3.9 million tonnes. The marine fish include pelagic and demersal stocks. Among pelagic resources, the major ones are oil sardines, anchovies, Bombay duck, ribbon fish and mackerel. Prawns, croakers, silver bellies, elasmobranchs, catfish, pomfrets and perch are the major components of the demersal stock. Estimates of the demersal resources in the deeper waters, particularly beyond 100 metres, are quite considerable, so these resources ought to be exploited.

E. Role of remote sensing for exploitation of ocean resources

Among the various oceanographic parameters, ocean colour and temperature are the most widely used in locating marine fishery resources. Temperature tells about the favourable environmental conditions, while ocean colour gives an indication of the standing stock of green biomass. Remote sensing from aircraft and satellites provides synoptic coverage. The development of quantitative techniques for the analysis of satellite data is contributing to the understanding of the spatial distribution of oceanographic parameters.

F. Remote sensing for primary productivity estimations

Phytoplankton are responsible for over 95 per cent of marine photosynthesis. As the dominant primary producers, they occupy key positions in marine food production and carbon dioxide regulation. Satellite-acquired data on ocean phenomena from the experimental Coastal Zone Colour Scanner (CZCS) and the operational AVHRR have provided the basic data for exploring living marine resources. The fishing industry has used data on sea surface temperature, surface currents and ocean colour to improve catch efficiency and enchance profitability. Satellite remote sensing of ocean colour is the only tool that can provide information on marine primary production on a global scale. It is also one of the few tools that allows oceanographers to synoptically assess phytoplankton distribution in regions of high spatial variability. Early calibration studies of the CZCS demonstrated their potential for using satellite measurements of ocean colour to determine the abundance and distribution of phytoplankton in the ocean. CZCS measurements have already improved understanding of the distribution of phytoplankton standing stocks and their temporal changes; such measurements also provide more realistic bases for interpolating from shipboard measurements to regional assessments.

G. Remote sensing of fishery resources

Satellite data is more economical than other sources of data. Satellite remote sensing data can assist maritime and oceanic developing nations in harvesting, conserving and mapping valuable fishery resources. Operational satellite applications include identifying fronts, upwellings and other areas of high productivity, providing timely data to local fishermen about potentially favourable zones, and predicting potentially lethal conditions such as oxygen depletion and pollution.

H. Location of fishing grounds

Locating potential fishing zones involves detection of oceanic features in the processed satellite image and subsequently transferring the selected thermal features on to the respective geo-reference map. Surface circulation features are important in defining marine fish habitats. These include the location and evolution of eddies and other circulation patterns. Cold or warm eddies are a major factor in determining favourable environmental conditions; their current boundaries are manifested by the thermal gradients. Optical and thermal characteristics of surface waters can be used as natural tracers of dynamic patterns. Sea surface temperature serves as a useful indicator of prevailing and changing environmental conditions. This is one of the important parameters that decides suitable environmental conditions for fish congregation. Fluctuations in SST are often the results of changes in other factors, such as currents, fronts, eddies and upwellings. Although the direct influence of temperature may be of limited significance except for species like tuna, it accounts indirectly for distribution of other species also.

I. Indian experience in identifying potential fishing zones

A national programme, titled Ocean Related Remote Sensing Applications, sponsored by the Department of Ocean Development, is being executed by the National Remote Sensing Agency, Hyderabad, in collaboration with other national institutes such as the Space Applications Centre, Ahmedabad; National Institute of Oceanography, Goa; CSIR Centre for Mathematical Modelling and Computer Simulation, Bangalore; Central Marine Fisheries Research Institute, Cochin; Institute for Ocean Management, Anna University, Madras; and Orissa Remote Sensing Application Centre, Bhubaneswar. Under this programme, SST over the Arabian Sea and Bay of Bengal is being retrieved using NOAA-AVHRR satellite data. Based on SST readings, thermal features such as eddies, current boundaries and thermal gradients are recognized and a forecast indicating the potential fishing zones is given regularly from Hyderabad.

J. Sea surface temperature retrieval

Sea surface temperature is retrieved from the thermal infra-red radiation from the sea surface and recorded by the AVHRR sensor of the polar orbiting NOAA satellite. Data on SSTs of the north Indian Ocean are available in the form of images. Figure 1 shows a weekly composite AVHRR-derived SST image over the north Indian Ocean.

K. Preparation of potential fishing zone maps

Daily SST images of three or four days are composited and the minimum and maximum temperatures are noted down; then these values are stretched from 0 to 255 grey levels in order to obtain maximum contrast of the thermal information (figure 2); from this relative thermal gradient image, features such as thermal boundaries, fine scale (0.2°C) relative temperature gradients to sharp (1°C) contour zones, fronts, eddies and upwelling zones are identified. These features are optically transferred on to corresponding sectors of the coastal maps prepared with the help of National Hydrographic Charts (figure 3). Major fish landing centres and seamarks are marked. Location of the PFZ with reference to particular fishing centres is drawn by identifying the nearest point of the thermal feature to that fishing centre. The information extracted consists of distance in kilometres, depth in metres (for position fixing) and bearing in degrees with reference to North for a particular fishing centre.

L. Potential fishing zone forecast dissemination

The PFZ maps are sent by facsimile transmission every Monday and Thursday to major fishermen associations, unions, and concerned government organizations such as CMFRI, FSI, MPEDA, ORSAC, MATSYAFED and state fishery departments of all the maritime states, including Andaman, Nicobar and Lakshadweep islands. The descriptive PFZ information is also sent by telex, telegram, telephone or NIC Net wherever fax facilities are not available. The concerned persons or organizations pass the PFZ information on to their members.

In addition, feedback forms are distributed to all the users. The forms are filled out and sent to NRSA regularly after each forecast. The feedback data consist of the date of fish catch, time of inhauling, latitude, longitude, name of the ship or boat, type of net used, depth of catch, distance away from the coast, direction, catch in kilograms, major catch in kilograms and major variety.

M. Potential fishing zone forecast validation

The fish abundance near the thermal fronts is fairly well established. Validation tests are being conducted using the data received at NRSA. In addition, scientists from NRSA undertake validation experiments, actually sailing to PFZ and non-PFZ regions. A substantial increase in fish catches was observed. The fishery forecasts provided to the fishermen along the Indian coasts were found to be fruitful for commercial exploitation, thereby saving fishing effort and fuel. It is planned ultimately to develop a fishery forecasting system in both the short-term and long-term by taking into account several biological and physico-chemical aspects. FSI fish catch data was validated extensively with the PFZ forecast data, and the fish catch in the PFZ areas was found to be more than three to four times as high as the catches in the non-PFZ areas. Validation results of the PFZ forecasts of Orissa, Andhra Pradesh, Tamil Nadu, Karnataka and Kerala states are given in figure 4; the results indicate, in all cases, more catch per unit effort (CPUE) in the PFZ areas than in the non-PFZ areas.

N. Future Indian programmes for monitoring ocean resources

With its recent experience, the Government of India has made plans (a) to launch an exclusive Ocean Remote Sensing Satellite for monitoring the Indian Ocean environment and (b) to further exploit ocean resources by using an ocean colour sensor, thermal sensor and microwave wind scatterometer by the end of 2000 AD.

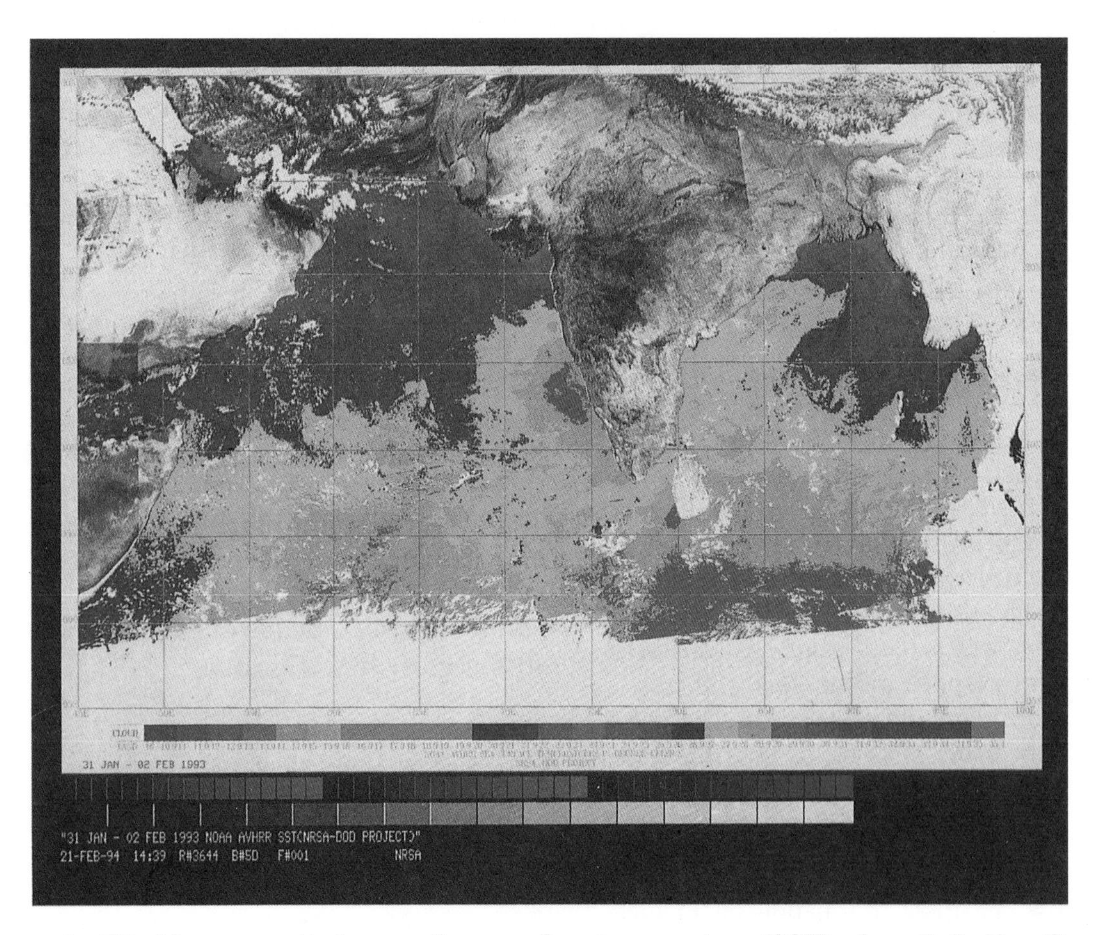

Figure 1. Weekly composite image of sea surface temperature (SST) of north Indian Ocean

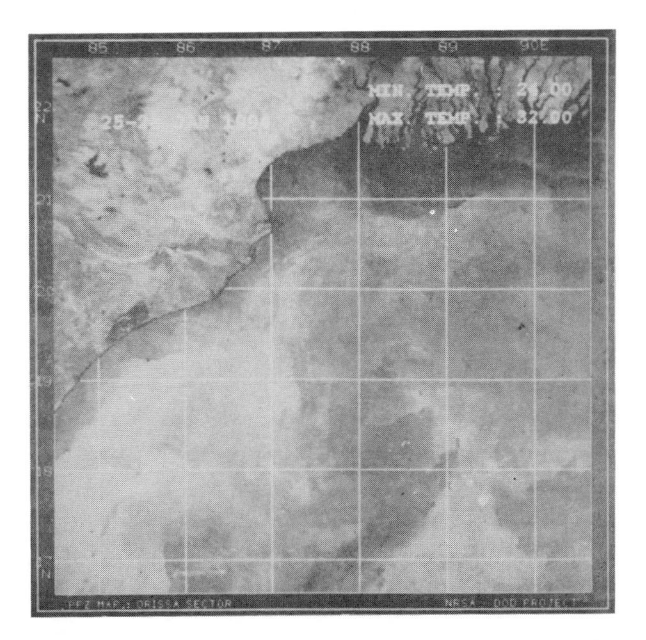

Figure 2. Image print of Orissa sector showing the thermal boundaries

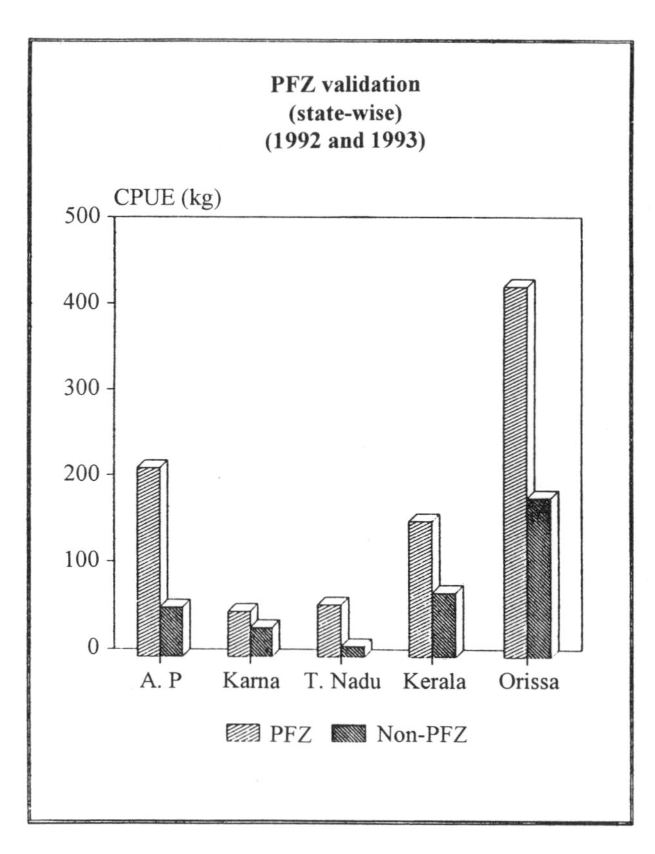

Figure 4. State-wise analysis of PFZ data validation chart

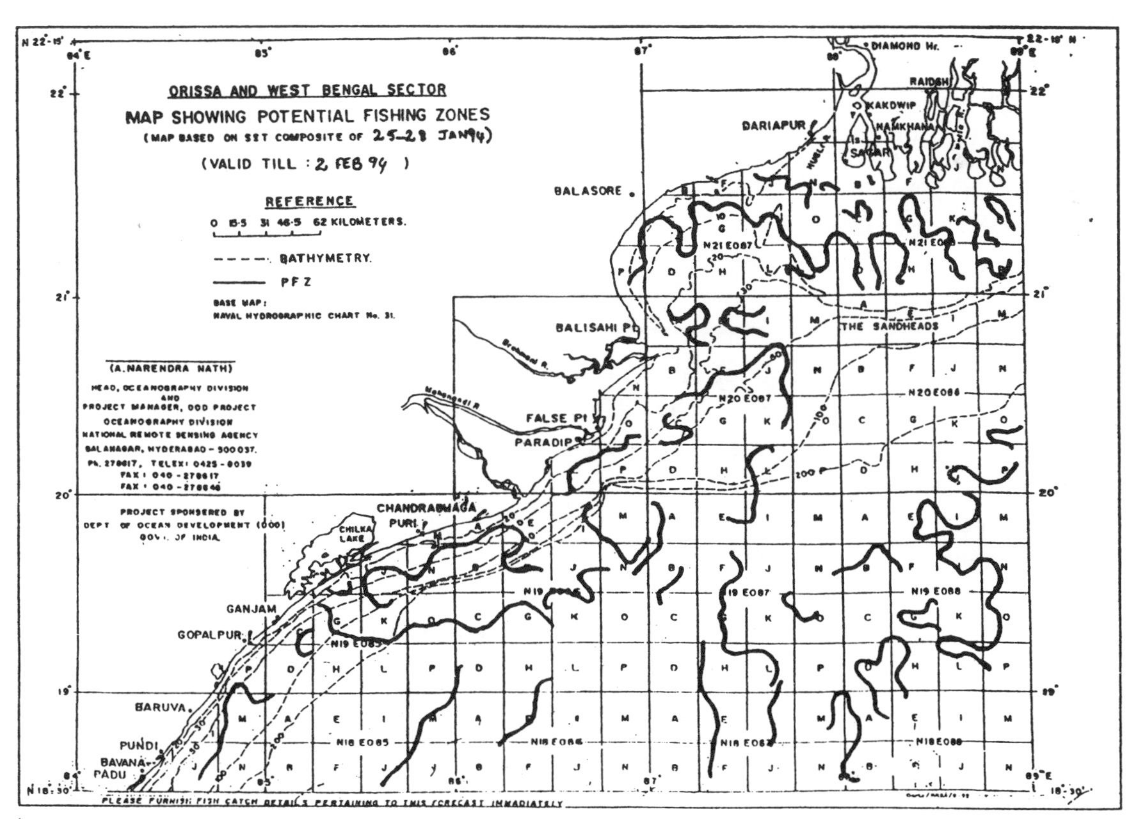

The boundaries and names shown and the designations used on this map do not imply official endorsement or acceptance by the United Nations.

Figure 3. PFZ map of Orissa sector showing the fish catch data

APPLICATION OF REMOTE SENSING IN GROUNDWATER STUDIES FOR RURAL DEVELOPMENT

*P.R. Reddy**

A. Introduction

Groundwater is one of the most important resources, in the context of sustainable rural development. It is essential for drinking as well as for irrigation purposes. In India more than 90 per cent of rural and nearly 30 per cent of urban population depend on ground water for meeting their drinking and domestic requirements. In addition, it accounts for nearly 60 per cent of the total irrigation potential in the country, irrigating about 32.5 million hectares, compared to 25.8 million hectares irrigated by all major and medium irrigation projects put together. Studies have proved that productivity from groundwater irrigation is almost double that of canal irrigation. Because of its longer residence time in the ground, low level of contamination, wide distribution and availability within the vicinity and within the reach of the consumer, groundwater development gets first priority for meeting the ever-growing demand for water in the rural areas. Therefore, it occupies an important place in the hydrologic cycle as well as in the life cycle of mankind.

B. Groundwater occurrence

The distribution of groundwater is not uniform in all the areas and it is subject to wide spatio-temporal variations. The movement and occurrence of water below the ground mainly depends on the aquifer characteristics of the underlying rock formations. The porosity and permeability of the rocks determine their water holding and transmitting characteristics (aquifer properties). Some of the rock formations, like unconsolidated/semi-consolidated sediments and volcanics, have primary porosity and permeability, whereas the consolidated rocks are hard, compact and devoid of primary porosity. However, even in hard rocks, secondary porosity may develop due to structural or tectonic forces at depth and weathering or decomposition of the rocks nearer the surface.

For proper understanding of the groundwater regime, information on different factors, i.e. the type of rock formations, their structural fabric, surface and subsurface expression, and source of water for recharge, are essential. Hence, to understand the movement and occurrence of water below the ground, the geology, geomorphology, structures and recharging conditions have to be well understood. If precise information on all these factors is known, it is possible to infer the groundwater regime, i.e. the type of aquifer, the suitable types of wells, the depth range, yield range, success rate area of influence of wells, and their sustainability, among other things. So it may be said that the groundwater regime in a given area is a function of lithology, land-form structures and recharge conditions prevailing in that area. However, depending on the situation, any of these factors may dominate over the others, exercising overall control on the system, or all of them may combinedly determine the groundwater regime.

C. Use of remote sensing data

The advent of remote sensing has opened up new vistas in geological, geomorphological and structural mapping vis-a-vis groundwater exploration. In the past, hydrogeological maps were prepared by conventional ground surveys based on the geological and hydrological information collected along traverse lines. While plotting such information on to the topographic base map and ultimately preparing the final maps by extrapolating the details, certain errors unavoidably used to creep in, leading to inaccuracies in the maps. Furthermore, it was difficult to collect integrated

* Geology Division, National Remote Sensing Agency, Hyderabad, India.

information on several factors that ultimately determined the groundwater regime. Hence, the geological community was in search of a suitable database for preparing accurate maps providing groundwater information.

Since the development of remote sensing technology, the mapping procedures have undergone a drastic change. Currently, the remote sensing techniques have assumed great significance and have become an integral part of the mapping programme, in both regional and detailed surveys. The scale-corrected satellite imagery in the form of either black and white or false colour composite (FCC) prints work as better databases for mapping. This pictorial database not only improves the accuracy of maps but also reduces the number of field traverses and observations to the bare minimum, saving time and cost of the investigations.

The synopticity of the satellite data helps in mapping of different lithologic, structural and morphological units in their correct spatial relationship, and the multispectral data provide certain additional information which is not easily observable on the ground. The repetitive coverage provides information on time-variant features and phenomena that are very important from a hydrologic point of view. Thus, the satellite imagery portrays an unbiased picture of the Earth's surface, providing integrated information on different factors controlling the groundwater regime and offering a common database for conjunctive study of all the factors for evaluating unit-wise groundwater potential prospects, utilization status and other factors.

D. Lithologic mapping

Lithology refers to an individual rock type, which is a basic geologic unit. The role of lithology in groundwater studies cannot be emphasized enough. Remote sensing provides the basis for discrimination and differentiation of rock types, depending on their spectral and morphological characteristics. Once the rock types are identified, the physical continuity of individual rock units can be easily traced and their exact shape, size and geometry can be identified and mapped with minimum ground surveys and more accurately by using satellite imagery. If areas have similar spectral and morphological characteristics, the interpretation can be extended or extrapolated to nearby areas also.

E. Structural mapping

As already discussed, in the hard rock areas the movement and occurrence of groundwater depends mainly on the secondary porosity and permeability resulting from structural discontinuities like lithologic contacts, bedding and cleavage planes, nonconformities, folds, faults, fractures and joints. Satellite imagery offers a better scope for understanding and mapping the geological structures both on regional and as well as detailed scales. It has the unique capability of providing direct evidence for regional-level structures like major faults, folds or lineaments, which cannot be directly observable on the ground because of the limited field of view. The synoptic view provides more accurate information about the form, configuration and continuity of different geological structures that generally appear discontinuous on the ground and, hence, cannot be mapped during field surveys because of the scarcity of rock outcrops, presence of soil cover and so on.

The most obvious structural features that are important from the point of groundwater are the lineaments. They occur as linear alignments of structural, lithologic, topographic, vegetational or drainage anomalies, either as straight lines or curvilinear features. They can be further classified into faults, fractures/joints, bedding traces, lithologic contacts, fold axes and so on based on their image characteristics and associations, in correlation with ground information.

The faults and fractures mainly act as conduits for groundwater movement; as such they form prospective groundwater zones, especially in the hard rock areas. The dykes and quartz reefs that occur as linear ridges or humps of low relief, in a discordant relationship to the country rocks, generally form barriers for movement of groundwater and compartmentalize the groundwater regime.

Often their linear trends are disturbed by slips or shifts due to cross-cutting faults. Based on their spectral character and morphologic expression, these dolerite dykes and quartz reefs can be easily identified and mapped using satellite imagery.

F. Geomorphological mapping

Geomorphology exercises significant control on the groundwater regime. The relief, slope, depth of weathering, type of weathered material, thickness of deposition, nature of the deposited material and the overall assemblage of different land-forms play important roles in defining the groundwater regime, especially in the hard rocks and the unconsolidated sediments.

Satellite imagery, thanks to its synoptic view, facilitates better appreciation of geomorphology and helps in identification and mapping of different land-forms and understanding their origin, sequence of evolution, material content and other traits. The satellite images, in combination with topographic maps and limited field checks at appropriate places, help in accurate geomorphic analysis and mapping of land-forms. While multispectral data give morphographic and morphometric details of different land-forms, the multitemporal data provide information on time variant features that indirectly indicate the origin, composition and material content of these land-forms.

Therefore, using satellite images, different erosional and depositional land-forms can be identified and grouped into different types, like continental, fluvial, alluvial, deltaic, marine, glacial, aeolian and others, depending on their origin. The palaeo-river courses, buried channels, alluvial valley fills, bajada deposits, closed valleys, deeply weathered pediplains, glacial tills, fracture/fault line valleys, karstic land-forms and so on form prospective groundwater zones, which can be easily identified and mapped using satellite data.

G. Hydrologic information and recharge to groundwater

The groundwater regime depends on the recharge conditions as the key input to the system. Without recharge there is no groundwater. The recharge to groundwater depends on many factors. By providing the integrated lithologic, structural, geomorphic and hydrologic data, satellite imagery facilitates better appreciation of the recharge conditions. Based on geomorphic and hydrologic analysis, the areas can be classified into runoff, recharge, storage and discharge zones, and the amount of recharge to groundwater from different sources can be assessed.

1. Mapping and monitoring of surface water bodies

Surface water bodies like reservoirs, tanks, lakes, canals and perennial stream/river courses form important sources of recharge to groundwater. Multi-date satellite data facilitate easy identification, mapping and monitoring of surface water bodies. Because of strong absorption in the near-infra-red region, the water bodies appear as dark patches on the satellite imagery, giving strong tonal contrast to the tone of the surrounding land areas. Therefore, by using satellite data, all water bodies can be accurately mapped both by visual and digital techniques and continuously monitored by multitemporal data. The surface water bodies, which act as recharge zones for groundwater, and the faults and fractures passing through these water bodies, which act as conduits for groundwater movement and carry water for long distances, can be easily mapped. Satellite imagery helps in identifying such inter-linkages, thus leading to better understanding of the groundwater regime.

2. Mapping of irrigated lands

Since approximately 25-40 per cent of the water used for irrigation goes as return flow to recharge the groundwater, the mapping of irrigated lands is useful in estimating the return flow component. Mapping of irrigated lands can be easily accomplished by using satellite imagery. Irrigated crops have maximum absorption in the 0.62-0.69 μ range owing to their high chlorophyll content, so they appear as dark patches on the satellite images. In standard FCCs, irrigated crops appear as bright red patches, also because of their high chlorophyll content. Therefore, owing to their strong absorption in the 0.62-0.69 μ range and high reflectance in the infra-red region (0.70-0.86 μ), the irrigated crops can be easily mapped both by visual interpretation and digital analysis techniques.

3. Groundwater draft estimation

Groundwater draft refers to the amount of groundwater being exploited in the area. As irrigation accounts for more than 90 per cent of groundwater utilization in the rural areas, the irrigated area statistics form one of the key inputs for groundwater draft estimation. The extent of the area irrigated by groundwater in different areas in different seasons is multiplied by per hectare consumption of water to get the gross groundwater draft in a given area. Satellite data helps in identifying, mapping and monitoring the areas irrigated by groundwater, i.e. other than canal and tank commands. Thus, it helps in understanding approximately the amount of groundwater that is being used during different seasons of the year in different zones.

H. Springs and groundwater discharge zones

Satellite imagery and aerial photographs help in locating groundwater discharge in the form of springs and seepages associated with marshy vegetation. Luxuriant growth of vegetation throughout the year also indirectly helps in identifying springs and seepages. Infra-red imagery is quite useful in identifying the marshy areas and moisture bearing zones associated with groundwater discharge.

I. Gaining and losing streams

Large scale enlargements of satellite imagery facilitate identification and mapping of streams or parts of stream channels where localized loss or gain occurs. Such information is useful in locating the gaining and losing reaches of streams, thereby giving an indication of groundwater movement in the near-surface environment. This information, viewed in combination with the geological conditions of that area, help in understanding the geological controls on groundwater movement.

J. GIS for data integration and map preparation

Information on all the factors discussed above can be put into a geographic information system as individual overlays, and then, by combining all the information, an integrated groundwater information system can be developed, which will provide information on all the aspects of groundwater regimes, which is essential for systematic planning for sustainable development of rural areas. Alternatively, integrated groundwater information maps can be generated manually by combining all the relevant information using appropriate symbols, annotation colour scheme and legend system. All this information can also be put into a set of maps ultimately to prepare a groundwater atlas of an area, which will be quite useful for effective planning and management of the groundwater resources.

K. Selection of well sites

In addition to the preparation of maps, satellite images at 1:50,000 to 1:12,500 scale are quite useful in preliminary selection of well sites. The large-scale satellite imagery provides clues for favourable lithologic, structural and geomorphic set-ups, where further detailed exploration can be taken up by well inventory and ground geophysical surveys.

The 1:250,000-scale images are useful in regional appraisal of the area and help in planning exploration programmes at the district level. On the other hand, the 1:50,000-scale geocoded products are quite useful to the field level scientists for tentative selection of sites. The large-scale data, at 1:25,000 to 1:12,500 scales, are useful in carrying out more detailed interpretation towards further narrowing down the target areas for ground hydrogeological and geophysical surveys, saving much time and effort.

L. Summary

Remote sensing data can be useful in many ways in groundwater exploration, such as preparation of regional and detailed maps, showing prospective groundwater zones, selection of well sites, groundwater resource estimation, draft estimation and calculating the balance, systematic planning and budgeting of groundwater, demarcation of over-developed areas, planning for joint use of surface and groundwater, selection of sites for artificial recharge, and other applications. Remote sensing data give best results when used as a complementary tool in connection with ground hydrogeologic and geophysical surveys.

PART EIGHT
OTHER APPLICATIONS

REMOTE SENSING AND GIS APPLICATIONS IN AZERBAIJAN

*Roustam Roustamov**

The Azerbaijan National Aerospace Agency (ANASA) is responsible for carrying out space research and its applications in different fields of the national economy.

Up to the end of 1991, Azerbaijan was part of the former Soviet Union and therefore it did not take part in any kind of international cooperation independently, including in the field of space research and its applications. Only after Azerbaijan gained its independence did it start to establish international links and cooperation. Azerbaijan hopes to take part in the ESCAP project in sustainable rural development using integrated remote sensing and GIS.

At present, ANASA includes a design bureau, pilot plant and three research institutes. The Space Research Institute of Natural Research deals mainly with the development of methods and techniques for remote sensing. At the Institute of Aerospace Information, the main scientific direction is the development of the software and hardware for aerospace data processing. The Institute of Ecology deals mainly with the ecological monitoring of the environment.

The work carried out in ANASA is focused in two main directions:

(a) Developing software for thematic interpretation of remote sensing data;
(b) Aerospace monitoring of the environment, including the Caspian Sea.

In terms of the first direction, ANASA is developing methodical, algorithmic programme software for the processing and thematic interpretation of remote sensing data for particular scientific and applied purposes.

The tasks of the second direction concentrate on the study of ecological conditions, development of GIS and the carrying out of regular aerospace monitoring of the territory of the country, with special emphasis on the Caspian Sea problem.

An important part of the work is developing GIS on the basis of united hardware and software, with which ANASA aims to:

(a) Collect and systematize many years of statistical data on natural resources and the environment;

(b) Create a subsystem for the processing and thematic interpretation of aerospace information;

(c) Create a subsystem of automated thematic mapping.

In order to establish the correlative dependence between the characteristics of the studied land objects and electromagnetic signals reflected or radiated by them, ANASA has carried out complete multi-stage remote measurements synchronous or quasi-synchronous with contact land measurements at specially chosen control test sites.

Control test sites are territories with a variety of objects both natural and man made, which could be representative of a bigger area. The synchronous experiments helped to solve different scientific methodological and practical problems in working out methods of remote sensing data interpretation. Also, contact land measurements of physical, biological and other characteristics of different objects at the test sites, as well as meteorological parameters, were used as a priori and/or control information.

* Azerbaijan National Aerospace Agency, Baku.

ANASA thus created a regional databank of spectral and agrometeorological information for calibration purposes. In order to unify and standardize the process of the software development, there was created an integrated program medium of processing which enables ANASA to perform the following work:

(a) The definition and evaluation of the condition and dynamics of water resources;

(b) The monitoring of water and soil pollution;

(c) The determination of the land-cover condition;

(d) Vegetation index monitoring, and the inspection of anthropogenic and technogenic influences on the environment;

(e) The study of the rapidly changing line of the Caspian Sea, and land degradation monitoring.

The development methods and algorithms of storage and processing of multichannel aerospace information make it possible to automate the process of cadastral survey of the land, which was of great importance in the period of land privatization that had started in Azerbaijan.

Using space and airborne images of the territory of Azerbaijan, along with developed processing methods, the following tasks have been executed:

(a) A map of tectonic tension of Azerbaijan to depths of 1.5-28.5 km was developed. The comparison of the developed map of tectonic tension with existing geological maps shows close correlation between them, thus proving the accuracy of the methods; these maps could also be used for discovering oil and gas structures;

(b) Maps were developed showing areas of mudflow danger. The mudflow on the mountain rivers of Azerbaijan are characterized by unpredic table outbreaks and cause serious problems for the population of quite large parts of the country. The newly developed maps of the southern part of the Caucasus enable ANASA to define the mudflow danger areas to determine the parameters of mudflow and evaluate the degree of mudflow danger for different regions. The maps may be used for all kinds of hydro-engineering work;

(c) A similar map of landslide danger areas was developed. The landslide danger was constant for some regions of the country;

(d) Maps of summer pastures, showing detailed parameters, were developed. The techniques and methods used for the developing of maps permit ANASA to provide regular remote sensing of the condition of the pastures and evaluation of their fodder potential. The development of such maps is explained by the huge significance of summer pasture for the country as a natural fodder base for sheep breeding;

(e) ANASA has also worked out the method of forecasting the yield of the cotton harvest, one of the main crops in Azerbaijan. The method is based on remote sensing data and a multichannel scanner. A great deficiency, however, is the lack of a ground receiving station.

A ground station located in Azerbaijan would enable ANASA to cover a huge territory from the eastern borders of Turkey -- the territory is at present not covered or covered only partly by acting ground stations, so help in discussion of the problem would help several countries of the region receive space-borne images and implement remote sensing methods in solving the problems of the national economy of those countries.

REMOTE SENSING AND GIS ACTIVITIES IN CAMBODIA

*Chuon Chanrithy**

The Cambodian people are working hard in the effort to develop the country and improve the living standard. To assist the people in this effort, the Royal Government is strengthening the legal framework and implementing various reforms in order to provide a fruitful environment for sustained economic growth. In the strategic plans, the development of human resources is considered a high priority in order to ensure the economic development in a sustainable manner, so the natural resources must be wisely and effectively used. To preserve nature for future generations, Cambodia recognizes the importance of environmental protection. Exploitation of natural resources and expansion of industries without appropriate environmental legislation and control would rapidly deteriorate the quality of life. The Royal Government is emphasizing sustainable development to ensure that the natural resources that provide the basis for growth are preserved for future generations.

The Ministry of Environment was set up to contribute to the protection, preservation, and conservation of the natural environment nationwide. To further this aim, provincial environmental offices have been deployed in 21 provinces under the supervision of the Ministry. Short and long-term strategy and planning of the Ministry include passage of a framework environmental law drafted in 1994; preparation of a national environmental action plan; drafting of regulations, guidelines and standards for environmental management implementation of the Cambodia Environmental Management Project; promotion of environmental education and awareness; and information/data collection for integrated information management, among other projects.

A. GIS and remote sensing overview

Recent developments in geographic information system (GIS) and remote sensing technology, brought about through the use of the computer and its related technology, have made it possible to compile geographical information, which, along with existing statistical data, have made the task of establishing fundamental references used in analysis more accurate and less time consuming. Under such conditions, the production, collection, accumulation and utilization of required information related to national development carried out using GIS will play a vital role in the establishment of a rational national development plan. This technology has become a very important tool for the management and monitoring of natural resources. An increasing amount and diversity of Earth observation data are becoming available as a new generation of sensor systems, particularly the radar satellites, are being launched and becoming operational. The information that can be derived from these data can contribute significantly to the knowledge of the natural resources of a country and assist in the formulation of development and management policies in a sustainable manner. For the profitable use of these rapidly evolving technologies, it is necessary to become familiar with the physical principles involved in the acquisition of data and with its use in specific applications.

GIS is increasingly being used for various environmental analyses. While this development has been helpful to planners, it is also important to assess the quality or accuracy of the results from such analyses because different software may use different methods or implementation procedures to calculate the required parameters. The use of GIS definitely facilitates the decision-making process and enhances the ability to develop proper plans. The result of GIS applied analysis would allow limited manpower and financial resources to be prioritized in solving the most urgent problems.

*Department of Socio-economic Resources and Environmental Data Management, Ministry of Environment, Phnom Penh, Cambodia.

B. Setting up of GIS office

To establish a GIS centre, an information system is essential. To better assess the preservation quality of a development project and to analyse and report on issues and concerns in any area, information has to be gathered, maintained and utilized by using various monitoring techniques supported by the use of GIS systems, which is why GIS is considered a tool for environmental study and analysis. The information system helps the government make informed decisions regarding sustainable development, and it provides an early warning system for environmental management. In this way, the use of GIS definitely facilitates the decision-making process and enhances the capability of planners to develop proper plans and prioritize manpower and finances in solving the most urgent problems.

Meanwhile, the Coastal and Marine Environment Management Information System Project (Phase I), jointly funded by the Swedish Government and the Asian Development Bank (ADB) as a regional technical assistance programme, has been implemented by the United Nations Environment Programme/Environment Assessment Programme for Asia and the Pacific (UNEP/EAP-AP) in conjunction with the Ministry of Environment.

Under the Coastal and Marine Environment Management Information System (COMEMIS) project implemented by UNEP/EAP-AP, the Ministry of Environment took this opportunity to set up its GIS office to meet the requirement. The data used through the computer-based information processing system can combine cartographic as well as tabular data and merge data from several sectors and sources for analysis. GIS is employed as an analytical tool, which can help subject matter experts as well as planners model, predict and model again until one or two clear options for action emerge. In short, it can help transform monitored data into useful information.

The objectives of the COMEMIS project are:

(a) To identify, assess, and collate the relevant and available information on the coastal and marine environment;

(b) To initiate and improve the collation, processing, storage, and assessment of the coastal and marine environment data and facilitate exchange and dissemination of information utilizing GIS;

(c) To improve the Ministry of Environment's capabilities in environmental planning by identifying and/or establishing relevant national databases and information systems;

(d) To improve networking and cooperation among the institutions utilizing GIS technology in the protection and management of the coastal and marine resources;

(e) To assist in building human and institutional capacities by providing training and computer hardware and software.

The project covered the training for the institutional staff capacity building and strengthening. For example, one staff of the Ministry was sent to the UNEP/EAP-AP office for internship in February 1995. Then, five staff were selected to attend the cycle-1 GIS training course conducted by experts from UNEP/EAP-AP held at the Asian Institute of Technology (AIT), Bangkok, Thailand, covering a 90-day period from May to July 1995. They built a database and worked together on the case study application to the coastal area of the country. The main activities for each of the thirteen weeks is shown below:

Week 1: Orientation, Computer Concepts, GIS Concepts, Raster View of GIS and Introduction to IDRISI.

Week 2: Data Retrieval and Display, Digital Elevation Models, Spatial Data Analysis and Overlay, Proximity Analysis, Map Algebra and Introduction to Remote Sensing.

Week 3: Vector View of GIS, Introduction to Arc Info, Database Design, Spatial Data Automation, Introduction to ADS, Digitizing with ADS, Digitizing ARCEDIT.

Week 4: Spatial Data Editing, Making Spatial Data Usable, Projection and Transformation, Attribute Data Entry, Introduction to Tables.

Week 5: Attribute Data Management, Database Query, Vector Spatial Analysis, Buffer Generation, Boundary Operations, Logical Operations, Spatial Joins, Cartographic Modeling, SML, Data Conversion.

Week 6: Map Design, Introduction to ARCPLOT, Interactive Map, Composition and Hard Copy Output, GPS, Windows, Arc View, Visit to NRCT.

Week 7: Project: Digitizing.

Week 8: Project: Digitizing.

Week 9: Project: Editing.

Week 10: Project: Editing.

Week 11: Project: Management.

Week 12: Project: Spatial Analysis, Map Output.

Week 13: Project: Map Output, Presentation and Graduation.

Afterwards came the stage of providing hardware/software and peripherals to the Ministry of Environment, in November 1995.

Hardware, software and peripherals provided by UNEP/EAP-AP include:

- PC Computer — 5 sets with GIS Arc Info software
- Digitizer — 1 set
- Plotter — 1 set
- Printer — 1 set
- UPS — 1 set
- Peripherals

The cycle-2 training was in-country, conducted by the participants from the cycle-1 training course with the assistance of the UNEP/EAP-AP experts as well. Five more staff of the Ministry of Environment were selected to attend in the training, which lasted over a two-month period from December 1995 to the end of February 1996. The training programme was chosen in accordance with the cycle-1 training with the case study application as well.

The last training course was the Advanced GIS training course, also conducted by UNEP/EAP-AP at AIT. Three technical staff of the Ministry were selected to attend the training, which lasted one month, in July 1996.

C. Future goal

The needs and requirements for an information system are never finished. The Ministry foresees the needs hereunder:

(a) The staff must have more training in GIS and remote sensing for different applications;

(b) Technical assistance from abroad has to be built, strengthened and developed;

(c) More GIS and remote sensing facilities are needed to facilitate the preparation of the development and management plans in a sustainable manner;

(d) Collaboration with relevant institutions such as Ministry of Agriculture, Forestry and Fishery; Ministry of Industry, Mine and Energy; Ministry of Public Works and Transportation; Ministry of Tourism; Ministry of Rural Development; Geographic Department; Land Titling Department; Cambodian Mine Action Centre; and international and regional agencies.

D. Conclusion

Because geographic information systems and remote sensing technology are very important tools for the management and monitoring of natural resources, the Royal Government of the Kingdom of Cambodia aims at using this tool to aid in rehabilitating the country and will pay strong attention to developing the country in a sustainable manner. In this matter, GIS and remote sensing technology is thought of as a vital tool for study and analysis, in that the Ministry of Environment has an important role to manage, improve and preserve the environment of the country. In order to meet the requirements for the successful application of remote sensing and GIS in sustainable development, the staff of the relevant institutions should be well trained. Cambodia still lacks experience and facilities in producing maps and is seeking technical assistance in order to improve the abilities of staff in institutions, enhance the availability and accessibility of environment and natural resource data, and establish an information exchange network and compatible data set for environmental management.

REMOTE SENSING AND GIS IN SRI LANKA

*Herath Manthrithilake**

A. Introduction

The tropical island of Sri Lanka is situated between 6° and 10° North latitude, near the equator. With an area of approximately 65,600 sq km, it is one of the smaller countries of the world. The island consists of central highlands rising up to more than 2,000 m, rounded by extensive lowlands. The land-forms range from undulating lowlands and wide flood plains to high plateaus, steep escarpments and urged mountains, thus forming a very attractive and interesting landscape.

While three-quarters of Sri Lanka belongs to the wet tropics, the northernmost part is characterized by a much drier climate. There is only a slight variation in the mean temperature throughout the year for most parts of the country, between 24° and 28° Celsius. There are two distinct monsoon seasons: yale and mala. Between them, the inter-monsoon periods provide rain throughout the country. In general, Sri Lanka is a beautiful country for many reasons.

B. Remote sensing and GIS

Remote sensing analysis started in early 1975 in Sri Lanka at the Centre for Remote Sensing, which is a division of the Department of Surveys, with assistance from Switzerland and technical advice from Zurich University. This project covered a very wide range of subject areas: development of land-use maps; paddy and yield estimation by cultivation season; existing and abandoned irrigation tanks and reservoirs; forest cover and crop cover extents; geological and soil surveys; groundwater assessment; pests and diseases; quality of crops; fishing grounds in the sea and island reservoirs; and weather forecasting. Some activities were carried out and successfully completed, while others were stopped at the pilot level.

Many human resource development activities were carried out with relevant line agencies. Unfortunately, none of these agencies established their capabilities or continued to work with the Centre for Remote Sensing as expected. Trained people were not retained on the job, or the trainers opted to go elsewhere, looking for greener pastures with the new skills they had acquired.

In 1987, the Forest Department started making inventories of their plantations and looked to remote sensing as a tool to update the island's forest-cover map, which was successfully completed. Hence, remote sensing continued to hold some interest in the area.

Meanwhile, a new focal point has emerged in remote sensing; that is the Arthur C. Clarke Centre for Modern Technologies, a government-affiliated, semi-independent centre. It has taken the initiative in satellite communication interests; however, this Centre too suffers from a drain of staff.

In the GIS sector there are more than 33 installations in the country, but many are not functioning. Although expensive software (like Arc Info) and hardware have been installed, these installations suffer from a lack of personnel experienced in reading and using the data. Many of the ones who remain are from various projects carried out by foreign consultants under foreign aid or loan programmes. Nevertheless, various institutions continue to install such equipment in their offices and departments at the Mahaweli Authority.

The Mahaweli Authority of Sri Lanka (MASL) is responsible for the country's largest multipurpose, trans-basin water resource development programme, based on the biggest river in the island: Mahaweli. Under this programme, 52 per cent of the country's electricity, 22 per cent of the

* Mahaweli Environmental Forest Conservation Division, Mahaweli Authority, Sri Lanka.

rice, 50 per cent of the onions, 70 per cent of the chillies and many other crops are being produced. Large numbers of people (132,000 families) are being resettled. Infrastructure and other socio-economic requirements were met. Nearly 35 per cent of the country will get direct benefits from this programme.

Under this programme, a large volume of data was gathered and continues to be gathered, so it was realized that geographic information systems (GIS) provided the best tool for understanding the meaning of this data. So in 1992 MASL stepped into GIS. To date, more than 5,000 sq km have been digitized at the 1:10,000 scale. Many other kinds of attribute data are also being linked to the spatial data base. Now MASL is working on human resource development and trying to retain people's expertise within the organization.

In this connection, MASL sees many benefits of remote sensing and GIS in the future, for the management of water, land and human resources, along with the other resources under its authority.

MASL is now planning to develop its remote sensing capabilities, so that it can rapidly update the situation on the ground, monitor new developments and predict and guide the farmers and development planners. In six months' time, MASL will install a PC-based receiving station for AVHRR information. It hopes to link and supply information to other agencies as well. We are looking forward to assistance from the National Remote Sensing Agency of India for developing this capability.

C. Issues

There are several problems that have been encountered in the development of remote sensing and GIS: (a) loss of trained staff to other countries and type of activities; (b) lack of constant focus on remote sensing/GIS development; (c) inaccessible expense of systems (software, data and hardware); and (d) difficulty of access to a multitude of procedures as well as data.

D. Conclusion

It is obvious that many developing countries could benefit from information and decision-support systems like remote sensing and GIS. Therefore, in Sri Lanka we look forward to more technical cooperation and assistance in this area.

APPENDIX: PARTICIPANTS

PARTICIPANTS

Azerbaijan

Mr. Roustamov Roustam Baloglan Oglu, Deputy General Director, Azerbaijan National Aerospace Agency, 159, Azadlyg pr., Baku 370106. Tel: (994-12) 62-17-22, Fax: (994-12) 62-17-22, Tlx: 142 218 SPACE AI, E-mail: mekhtiev@anasa.baku.a3

Bangladesh

Mr. Hafizur Rahman, Senior Scientific Officer, Bangladesh Space Research and Remote Sensing Organization (SPARRSO), Agargaon, Sher-E-Bangla Nagar, Dhaka 1207. Tel: (88-02) 327913, Fax: (880-2) 813080

Cambodia

Mr. Teng Peng Seang, Chief, GIS Division, Integrated Resource Information Centre (IRIC), 200 Norodom Boulevard, Phnom Penh. Tel: (855-23) 725107, Fax: (855-23) 725007, E-mail: iric@forum.org.kh

Mr. Chanrithy Chuon, Deputy Director, Department of Socio-economic Resources and Environmental Data Management and GIS Specialist, GIS/Remote Sensing Office, Ministry of Environment, 48 Samdech Preah Sihanouk Avenue, Chamkarmon, Phnom Penh. Tel: (855-15) 921020, Fax: (855-23) 427844

China

Mr. Tang Hua Jun, Director-General, Institute of Natural Resources and Regional Planning, Chinese Academy of Agricultural Sciences, 100081-Beijing. Tel: (0086-10) 62187416, Fax: (0086-10) 62174142, E-mail: hjtang@sun.jhep.ac.cn

India

Mr. D.P. Rao, Associate Director, National Remote Sensing Agency, Balanagar, Hyderabad 500 037. Tel: 279677, Fax: 040-279677

Mr. T. Ch. Malleshwar Rao, Group Head, Image Analysis and Inter/Applications, National Remote Sensing Agency, Balanagar, Hyderabad 500 037. Tel: 279572 Ext. 2229, Fax: 040-278648

Mr. Mukund Rao K., Manager, U and US, NNRMS, ISRO, Headquarters, Department of Space, Antariksh Bhavan, New B.E.L. Road, Bangalore 560 094. Tel: 3416413, Fax: 080-3412471

Mr. P.S. Roy, Head, Forestry and Ecology Division, Indian Institute of Remote Sensing, No. 4, Kalidas Road, Post Box No. 135, Dehradun 248 001. Tel: 0135-654518, Fax: 91-135-651987

Mr. A.T. Jayaseelan, In charge, Water Conservation Section/WRG, National Remote Sensing Agency, Balanagar, Hyderabad 500 037. Tel: 279572 Ext. 2205, Fax: 040-278648

Mr. J.S. Parihar, Head, ARD/RSAG, Space Applications Centre, Jodhpur Tekra, SAC Post, Ahmedabad 380 053. Tel: 447043, Fax: 079-462677

Mr. M.V. Krishna Rao, Head, Agriculture Division/Agriculture and Soils Group, National Remote Sensing Agency, Balanagar, Hyderabad 500 037. Tel: 279572 Ext. 2264, Fax: 040-278648

Mr. B.R.M. Rao, Head, SS and LED/Agriculture and Soils Group, National Remote Sensing Agency, Balanagar, Hyderabad 500 037. Tel: 279572 Ext. 2257, Fax: 040-278648

Mr. R.S Dwivedi, Head, LDD/Agriculture and Soils Group, National Remote Sensing Agency, Balanagar, Hyderabad 500 037. Tel: 279572 Ext. 2227, Fax: 040-278648

Mr. T.P. Singh, Scientist, Space Applications Centre, Jodhpur Tekra, SAC Post, Ahmedabad 380 053. Tel: 447043, Fax: 079-6568073

Mr. P.R. Reddy, Head, Geology Division/Geo. Group, National Remote Sensing Agency, Balanagar, Hyderabad 500 037. Tel: 279572 Ext. 2262, Fax: 040-278648

Mr. Y.V.N. Krishnamurthy, Deputy Head, Regional Remote Sensing Service Centre (RRSSC), NBSS and LUP Campus, Shankarnagar Post, Amaravathi Road, Nagpur 440 010. Tel: 0712-531393

Mr. S.K. Subramaniam, Scientist/Engineer, National Remote Sensing Agency, Balanagar, Hyderabad 500 037. Tel: 279572 Ext. 2213, Fax: 040-278648

Mr. A.N. Nath, Group Head, Oceanography/Applications, National Remote Sensing Agency, Balanagar, Hyderabad 500 037. Tel: 278617, Fax: 040-278648

Mr. R.S. Rao, Director, Andhra Pradesh Remote Sensing Application Centre (APSRAC), Nagarjuna Hills, Punjagutta, Hyderabad. Tel: 319761, 393327, Fax: 040-316254,

Shri Anil Kumar, Group Head, Earth Station Operations/DA-S, National Remote Sensing Agency, Balanagar, Hyderabad 500 037. Tel: 278009, Fax: 040-278648

Mr. C.B.S. Dutt, Head, Forestry Division/Application, Group, National Remote Sensing Agency, Balanagar, Hyderabad 500 037. Tel: 279572 Ext. 2266, Fax: 040-278648

Indonesia

Mr. Iman Bonila Sombu, Expert Staff, Directorate General of Rural Community Development, Ministry of Home Affairs, Jalan Raya Pasar Minggu Km. 19, Jakarta-Selatan. Tel: (021) 7995104, Fax: (6221) 7941939

Mr. Kawit Widodo, Head of Sub-directorate for Rural Area Improvement, Directorate General of Rural Community Development (DGRCD), Jalan Raya Pasar Minggu Km. 19, Jakarta-Selatan. Tel: (021) 7942373, Fax: (6221) 7942517

Islamic Republic of Iran

Mr. Mohammad Reza Varasteh, Head, GIS Affairs of Iranian Remote Sensing Centre, 22-14th Street Saadat Abad Avenue, Tehran. P.O. Box 1136516713, Tel: (021) 2064471-3, Fax: (021) 2064474

Republic of Korea

Mr. Keun-Seop Shim, Researcher, Statistical Analysis Division, Farm Management Bureau, Rural Development Administration, Suwon. Tel: (82-331) 2924162, Fax: (82-331) 296-2328

Malaysia

Mr. Loh Kok Fook, Head, Applications and Image Processing Division, Malaysian Centre for Remote Sensing (MACRES), Letter Box 208, CB 100, 5th Floor, City Square Centre, Jalan Tun Razak, 50400 Kuala Lumpur. Tel: (093) 2645640, Fax: (603) 2645646, E-mail: maures@maures.saims.my

Mongolia

Ms. Baghaday Naranchimeg, GIS and Land Resources Information Section, Research Institute for Land Policy, Chingunjav, Street 2, Ulaanbaatar-35. Tel: (976-1) 360471, Fax: (976-1) 360506

Myanmar

U Aung Kyaw Myint, Associate Professor, Institute of Foresty, Forestry Institute, Yezin, Pyinmana Township, Mandalay Division. Tel: (095-67) 21574, (095-01) 641462, Fax: (095-01) 664336

Nepal

Mr. Pradip P. Upadhyay, Section Officer, National Planning Commission Secretariat, Singha Durbar, Kathmandu. Tel: (977-1) 225879, Fax: (977-1) 226500

Mr. Gyani Babu Juwa, Remote Sensing Officer, Section Head, Forest Research and Survey Centre, Babarmahal, Kathmandu. Tel: (977-1) 222601, Fax: (977-1) 226944

Sri Lanka

Mr. Herath Manthrithilake, Director, Mahaweli Environment and Forest Conservation Division, Mahaweli Authority, Dam Site, Polgolla. Tel: (94-078) 70600, Fax: (0094-8) 234950, E-mail: efcdmasl@slt.lk

Thailand

Mr. Suthep Chutiratanaphan, Agriculturist, Land Use Planning Division, Land Development Department, Phaholyothin Road, Chatuchak, Bangkok 10900. Tel: (662) 5791937, Fax: (662) 5797589

Viet Nam

Ms. Tran Thi Kieu Hanh, Scientific Researcher, Institute of Geography, Viet Nam National Centre for Science and Technology, Nghia Do Tuliem, Hanoi. Tel: (84-4) 8362607, Fax: (84-4) 8361192

Observers

Mr. Rajendra Kumar Gupta, Group Head, Training and Educational Activities, National Remote Sensing Agency, Balanagar, Hyderabad 500 037. Tel: (040) 278870, Fax: 040-278648, Tlx: 0425-8039, E-mail: rkg.nrsa-hyd@uunet.in

Mr. S. Prasad, Scientist, Training and Educational Activities Group, National Remote Sensing Agency, Balanagar, Hyderabad 500 037. Tel: 040-278870, Fax: 040-278648, Telex: 0425-8039, E-mail: rkg.nrsa-hyd@uunet.in

Mr. P.V. Krishna Rao, Scientist, Agriculture and Soils Group, National Remote Sensing Agency, Balanagar, Hyderabad 500 037. Tel: 040-279572 Ext. 2263, Fax: 040-278648

Mr. G.Ch. Chennaiah, Scientist, Integrated Survey Group, National Remote Sensing Agency, Balanagar, Hyderabad 500 037. Tel: 040-279572 Ext. 2213, Fax: 040-278648

Mr. Y.V.S. Murthy, Scientist, Integrated Survey Group, National Remote Sensing Agency, Balanagar, Hyderabad 500 037. Tel: 040-279572 Ext. 2213, Fax: 040-278648

Mr. T.S. Prasad, Scientist, Training and Educational Activities Group, National Remote Sensing Agency, Balanagar, Hyderabad 500 037. Tel: 040-278870, Fax: 040-278648, E-mail: rkg.nrsa-hyd @uunet.in

ESCAP Secretariat

Ms. Claire Gosselin, GIS/Remote Sensing Specialist, ESCAP, United Nations Building, Rajadamnern Nok Avenue, Bangkok 10200, Thailand. Tel: (662) 288-1415, Fax: (662) 288-1000, E-mail: gosselin. unescap@un.org

Mr. Tsutomu Shigeta, Expert on space technology applications, ESCAP, United Nations Building, Rajadamnern Nok Avenue, Bangkok 10200, Thailand. Tel: (662) 288-1458, Fax: (662) 288-1000, E-mail: shigeta@un.org